U.S. Food: Making the Most of a Global Resource

Other Titles of Interest

Food From the Sea: The Economics and Politics of Ocean Fisheries, Frederick W. Bell

Protein, Calories, and Economic Development: Nutritional Variables in the Economics of Developing Countries, Bernard A. Schmitt

A Westview Special Study

U.S. Food: Making the
Most of a Global Resource
David W. McClintock

This volume focuses on the basic dilemma inherent in the formulation and implementation of agricultural policies in the United States: many of the best short-term options are the worst ones for the long run, and vice versa. The study begins with an overview of the world food problem, including both the negative and positive issues that give rise to conflicting pessimistic and optimistic predictions regarding long-term food availability. After analyzing climatic and population variables and their effects on the prediction process, Dr. McClintock addresses a critical question: what are the real dimensions of U.S. agricultural "advantage," and how can this advantage be rationally used to accomplish often incompatible humanitarian, economic, and political objectives? He concludes with a look at the practical problems involved in the development of U.S. food policies in the next two decades and suggests areas to which government and public attention should be directed.

David W. McClintock is coordinator for political studies at the Foreign Service Institute in Washington and coordinator of the Foreign Affairs Interdepartmental Seminar. He received a Ph.D. from the University of Michigan and has been a U.S. Foreign Service officer for the past nineteen years, serving in the Philippines, Lebanon, Yemen, Jordan, and Washington.

U.S. Food: Making the Most of a Global Resource

David W. McClintock

Published in cooperation with
the National Defense University

Routledge
Taylor & Francis Group

LONDON AND NEW YORK

First published 1978 by Westview Press

Published 2018 by Routledge
52 Vanderbilt Avenue, New York, NY 10017
2 Park Square, Milton Park, Abingdon, Oxon OX14 4RN

Routledge is an imprint of the Taylor & Francis Group, an informa business

Copyright © 1978 by Taylor & Francis

Library of Congress Cataloging in Publication Data
McClintock, David W.
    U.S. food, making the most of a global resource.
    (A Westview special study)
    Bibliography: p.
    1. Food industry and trade—United States. 2. Agriculture—Economic aspects—United States. 3. Agriculture and state—United States. 4. Food supply. I. Title.
HD9005.M17   338.1'9'73   78-17302

ISBN 13: 978-0-367-21239-1 (hbk)
ISBN 13: 978-0-367-21520-0 (pbk)

Contents

Tables and Figures

U.S. Food: Making the Most of a Global Resource

1
The Many Faces of the Global Food Problem

Many imponderable problems challenge the development of a comprehensive, forward-looking U.S. food policy. Yet the development of such a policy is not an impossible task if appropriate efforts are made to develop linkages between seemingly disparate issues. The following chapters explore critical climatic and population variables and suggest possible ways to rationalize the humanitarian, commercial, and political uses of U.S. food as a national asset and a global resource.

The Challenges to U.S. Food Policymakers

Many food policy dilemmas will confront U.S. decisionmakers in the coming years. These dilemmas should lend further credence to the Club of Rome thesis that "The emerging world system requires a 'holistic' view to be taken of the future world development: everything seems to depend on everything else."[1] There is always a tendency, fueled by immediate needs, to seek short-term gains at the expense of long-range benefits. Grave consequences can result if a satisfactory short-range situation is mistaken for a long-range one. It appears that U.S. leadership faces a double challenge: to develop sufficiently far-reaching food and agricultural policies to maximize both national and international humanitarian interests, and to devise such policies in spite of inevitable pressures from domestic and foreign constituencies for less than satisfactory mini-solutions that would serve only their own interests.

Although the final quarter of the twentieth century has only just begun, one can safely predict that it will have global problems of a markedly different character from those of the post-World War II era. East-West competition for influence in the developing world, originally a manifestation of U.S.-Soviet rivalry in other spheres, soon may be overshadowed by North-South tensions between nations at various stages of economic development. Distinctions are now being made among the Third, Fourth, and even Fifth Worlds. From the U.S. viewpoint, this change in the global environment began about 1973, when Soviet grain purchases and the Arab-Israeli war provided highly visible evidence of the respective roles of food and energy in international economics and politics. The rudest shock came from the petroleum boycott by the Organization of Petroleum Exporting Countries (OPEC) and the resultant price increases. The boycott showed that the United States was more vulnerable than it had anticipated in terms of fuel imports; the grain sales showed that an exportable and frequently surplus domestic product—food—had become a new element in Soviet-U.S. relations as well as a more valuable item in the world marketplace.

A great deal of national concern, if not remedial action, has been focused on the energy problem, but comparatively little attention has been devoted to the question of what our role would be if there were a global food crisis resulting from unchecked population growth and inadequate food production. The varied and, at times, contradictory nature of recent private and government pronouncements on the subject suggests that food will be a more complex issue than energy by the end of the century. Some proponents of strategic food planning argue for more effective use of food as a "weapon" against the Soviets, uncooperative foreign oil exporters, and other international adversaries. Others in private and government circles decry the use of food as a weapon and, instead, point to the moral problems inherent in a situation where the United States may have a partial voice in determining who will eat and who will starve. The word "triage," describing the World War I concept of dividing battlefield casualties into three groups according to survival prospects, has been revived in some futurists' discussions of worst-case food scenarios.

The issue of future management of U.S. food becomes even more complicated when one considers the strong disagreements about the actual capacity of the U.S. agricultural system if simultaneously subjected. to domestic and global population pressures. There is also disagreement over the demographic predictions themselves. In addition, controversy has arisen over whether various predicted climatic changes—ranging from a new cooling or warming cycle to increased variability or perhaps another Midwestern drought—will have significant impact on the food situation and whether the U.S. position as a net food exporter will be strengthened or weakened as a result. Pessimists argue that climate, demography, and the predictable action (or inaction) of various national and international entities will lead to mass starvation. They point to the fact that U.S. agriculture is energy-intensive; they say that food and energy are not divisible issues but, rather, components of a macroproblem resulting from the increasing interdependence of all nations. The optimists foresee scientific breakthroughs extending the "green revolution,"* the cultivation of marginal and fallow lands, ocean harvesting, and the development of synthetic foods as solutions that will thwart the arrival of the ultimate crisis.

Innovative food policies for both the short- and long-term will be essential to preserve whatever advantages the United States derives from its remarkably productive agricultural system. Policymakers must face the additional challenges of reconciling conflicting farmer and consumer interests in the domestic

---

*Green revolution" refers to a series of highly successful hybridization experiments conducted primarily in Mexico and the Philippines that led to the development of hardy, productive seed grains for underdeveloped tropical and semitropical regions.

political arena, of balancing humanitarian requirements with political strategy in the international sphere, and of dealing with new fiscal realities and a host of other issues. The distant future is as unpredictable as ever. And it is difficult to plan for the short-term in a substantive area where there is so much disagreement on both technical and ethical grounds.

Dimensions of the Problem

The alarmists' arguments are formidable. As food expert Lester Brown observes, the annual increase in global demand for cereals (resulting primarily from population growth) is currently in excess of 30 million tons per year. In 1900 the annual increase was only 4 million tons. Half a century later, the figure had increased only threefold to 12 million.[2] Using the United Nations' "medium" projection of a world population of 6.5 billion by the end of this century, and assuming an average food consumption level approximating that currently prevailing in Western Europe (1,000 pounds of grain per person per year), world cereal production would have to reach almost 3 billion tons annually by the year 2000, which is 2.5 times the current output.[3] World population expansion alone, without any rise in per capita food intake, would require an increased production of three-quarters of a billion tons by A.D. 2000. This represents 2.5 times the current total production of North America. At the (unrealistic) U.S. level of consumption (875 kilograms record; 700 kilograms currently), the global requirement would reach 4.6 to 5.7 billion tons. According to the International Food Policy Research Institute, the continuation of global grain production at levels recorded over the past fifteen years would find the developing countries with a deficit of about 100 million tons per year as early as 1985–86. If the lower annual rate of increase in production characteristic of the late 1960s and early 1970s prevails instead, their annual deficit could reach 200 million tons.[4] As the situation in India and the highly publicized mid-1970s famine in the Sahel demonstrate, food problems can be endemic. Ironically in 1975 and 1976 India was faced with food supplies that exceeded handling capacity, falling agricultural prices, and a sharp increase in wastage.

Considering the position of the United States as an agricultural (as well as industrial and military) superpower, the implications of these predictions are more profound than policymakers, who must deal with the recurrent short-term problems of domestic production and overseas trade, might assume. Far-reaching decisions will have to be made with regard to production and pricing, storage, overseas marketing, and foreign aid. Policymakers will have to take into consideration the new and unfamiliar uses of food as a strategic commodity in an era of dwindling energy and raw material resources. Significantly, U.S. grain exports accounted for one-half or more of the world total in the period of 1973-75. The equivalent of 20 percent of all U.S. farm acreage is devoted to export production, which in turn occupies about 14 percent of the farm work force.[5] Since 1968 the value of U.S. agricultural exports has risen markedly (271 percent; from $6.2 billion in 1968 to $23 billion in 1976).[6] Notwithstanding these impressive figures and the obvious advantages that the United States derives from its land and technological resources, the national productive capacity is now the subject of controversy. It has been argued that even the term *surplus* is a relative concept not to be confused with nonmarketability. Tillage vastly increased in the early 1970s as a result of high market prices. In some instances it increased at the risk of creating new dust bowls if the climatic conditions of the 1930s return. Although there is a lot of fallow land in the United States, much of it is marginal in terms of the efficiencies of mechanization.[7] Some observers warn that the United States cannot expect the windfall in production increases that resulted from the earlier revolution in mechanization. Pessimists question whether present nutritional levels can be upheld, disregarding the demand for improved diets likely to result from increased per capita income. Increased agricultural production would be required from an acreage which in all probability will have shrunk as a result of urban and industrial expansion. Georg

Borgstrom discusses some of the implications of reduced acreage:

> By then the yields per acre will have to be increased considerably,
> demanding major investments and large-scale engineering projects
> for irrigation, fertilizer plants, and sewage disposal units of en-
> tirely new types. . . . [Faster growing cycles] will necessitate a
> much higher consumption of fertilizers (by some experts the U.S.
> requirements have been estimated as 17 times higher by the year
> 2000 than in 1969), and, above all, water. . . . Water tables in
> several key areas already have fallen below the level for eco-
> nomically acceptable pumping.[8]

On a more positive note, one should not underestimate the
political and economic advantages that the scale and technological development of our agriculture affords our country.
Indeed, doubts concerning the long-term productive capacity
of the American farmer are difficult to comprehend considering
renewed surpluses and weakened prices.

There has been a temptation to regard the American agricultural bounty in somewhat simplistic terms, particularly in terms
of food as a political commodity. For example, it must be
remembered that the recurring Soviet demand for U.S. grain
results from a conscious political decision on their part to adhere
to livestock development plans despite domestic grain shortfall.
It does not come from an appraisal that direct human consumption of cereals will have to be reduced. Improving the quality of
the Soviet diet has become an important political commitment
of the Kremlin leadership, but they regard U.S. grain purchases
as an option rather than as a necessity. The situation with
regard to the petroleum-exporting nations, likewise, is not ideal
for application of a U.S. *food lever*. Because of their enormous
revenues and fairly modest food import needs (based on their
small populations), most OPEC nations could obtain from the
international marketplace any foodstuffs that the United States
withheld. (The only truly populous OPEC nation is Indonesia,
which has its own fairly respectable agricultural base. Nigeria

is another possible exception.) Similarly, U.S. ability to feed its European (or other) allies in the event of a widespread military conflict is limited. It should be judged in terms of supplementary assistance rather than as the large-scale feeding operation sometimes hypothesized in war-games. At the production levels of the mid-1960s, one-half of all U.S. exports of fuel and feed—the equivalent of 40 million acres—would have been required to sustain England at its customary dietary level; Western Europe could not be sustained entirely by U.S. agriculture.[9]

The Aid Dilemma

The extent to which food may be used for humanitarian assistance is limited by the unfavorable ratio of need to availability and by cost realities. Distinctions must be drawn between famine relief in low-population areas, such as the Sahel, and famine relief in heavily populated areas of periodic need, such as India. Food aid programs only occasionally represent a single policy objective somewhere between pure humanitarianism and pure pragmatism. Just as the Soviet deals of the early 1970s were motivated by both politics and economics, our aid to the Third World has been motivated by both altruism and enlightened political self-interest.

Food is produced as the result of conscious decisions about planting and other resource allocations, is perishable after it is produced, and involves complex costs in handling, storage, and transportation. These factors do not preclude its use either as a weapon or as a purely humanitarian resource. They underscore the fact that practically every national policy decision relating to food involves trade-offs and complex constituency pressures. U.S. policies toward food aid obviously will require periodic reassessment in the light of adverse shifts in global food balance and emergence of new, competing domestic budgetary priorities. New cost calculations will have to be weighed. The United States' instrumental role in structuring the global food balance must be assessed in terms of implied future responsibilities as well as current political advantages.

In a publication entitled *The World Food Situation and Prospects to 1985,* the U.S. Department of Agriculture (USDA) notes that the total value of U.S. food aid programs has remained fairly constant, around $1 billion per annum in the 1970s, but the *volume* of this aid has declined as both the volume and value of exports has risen. Indeed, the volume of commodities delivered under U.S. food aid programs by 1974 was the lowest since the mid-1950s, when shipments began under Public Law 480.[10] In previous years the use of surplus food as a component of foreign aid helped reduce storage costs. This in effect made food a cheaper aid item than capital goods or services obtainable only at market prices.

Although the United States has been encouraging other nations to share the international burdens, the results have not been encouraging. Despite improvements in world agricultural production in the 1975–76 crop years, earlier poor harvests in Europe and shortfalls in the USSR limited the net global production increase to only 3 percent in 1976, an amount that was not comfortably ahead of the world population increase of 1.9 percent. The now-defunct Subcommittee on International Resources, Food, and Energy of the House International Relations Committee generated considerable pressure for larger U.S. aid commitments with its 1976 Right to Food Resolution. The resolution promoted self-help development among the world's poorest countries, especially those most seriously affected by hunger, and emphasized the plight of the rural poor. Significantly, it set a target of 1 percent of the U.S. gross national product (GNP) to be spent on all forms of U.S. food assistance, governmental and private.[11] Although the resolution passed in both the House and Senate in separate versions, the two chambers could not agree on a consolidated version before expiration of the session. In 1977, congressional attention shifted to a World Food Reserve Bill. Even though the "Right-to-Food" hearings evoked a favorable response, it is doubtful that U.S. grants and concessional sales of food will ever return to the levels witnessed when surpluses were a major domestic agricultural problem. And it is questionable whether, over the long-term, the United States will find it prudent or feasible to hold to fixed aid commitments (based on percentage of GNP) for a commodity that is becoming a critical factor in the international balance of payments.

The aid dilemma cannot be resolved solely by determining how much food will be exported through gifts or loans from the United States and other food-exporting nations. Other developmental problems must be resolved so that the deficit countries can increase their own agricultural production. Without this parallel effort, international aid will inevitably fall short of meeting even minimal actual needs. For example, despite the deficit situation prevailing in many developing countries, the influx of foreign food aid tends to hold down local agricultural prices. Foreign farmers, who also fear the effect of periodic gluts and surpluses on prices, thus are discouraged from expanding their production. This tendency may be further reinforced by local government policies that hold market prices at artificially low levels as an indirect subsidy to urban populations. Ironically (as argued in one study by Bridger and de Soissons), world food production cannot be increased by more than 3 to 4 percent per year for any sustained period without incurring the risk of unsalable surpluses, particularly in the surplus-exporting countries like the United States. This problem diminishes only when increased efficiency leads to sufficiently lower costs, reflected at the retail level. But this, in turn, can create a demand for higher quality food from an entirely new income sector; more often than not, other sectors' incomes will not rise sufficiently to absorb the costs of increased agricultural production at this 4 percent level.[12] If aid to the industrial sector is too intensive, industry may have to draw from the rural population. This possibility has the multiple negative effects of reducing agricultural manpower, increasing pressures on the local government for urban subsidies, and adding to the donor nations' burdens in the food sector. A cruel fact of life is that premature or overzealous health programs, in the absence of companion developmental efforts, tend to promote hunger and malnutrition.

From the foregoing, it is apparent that the United States must balance its future roles as both aid donor and commercial food exporter such that the burdens of donorship can eventually be exchanged for the benefits of export trade. One of the essential requirements in this regard is to determine the actual needs of potential aid recipients; market forces, of course, will determine commercial supply and demand.

Measuring Needs

In determining the requirements of the poorer deficit nations, one finds that a frequently overlooked piece of information relates to calorie and protein levels. There is a diversity of opinion about quantity and about the specific types of food needed to carry a population over the threshold from malnutrition and undernourishment to the health of a sufficient and balanced diet. The latter determination is an important concern because other developmental efforts simply will not work in the absence of physical and mental vitality on the part of the participants.

Both dietary balance and the amounts of food needed for physical fitness depend on such factors as sex, age, body weight at different ages, temperatures at which people live, and the work they do. The actual effects of temperature are subject to controversy. Basically, less food is required to maintain body temperature in a warm climate than in a cold one; accordingly, inhabitants of cooler climatic zones need more food bulk. However, there is evidence that the energy cost of hard manual work in hot climates—where most of the malnourished people live—is 15 percent higher than average.[13] Increased crop yields from the so-called green revolution have altered ratios among the bulk, protein, and calorie components of the new crops. The task of projecting global food production and consumption has therefore become more sophisticated.

The problem of identifying trouble spots is comparable to

the difficulty of determining basic dietary requirements. Although there is an important linkage between regional famine and global instability, in the opinion of food expert Lester Brown, the world's affluent regions are able to ignore impending tragedies because the way famine manifests itself has changed. "In earlier periods, famine was largely a geographical phenomenon," he says; " . . . Famine today tends to be spread more evenly among the world's poor rather than concentrated in specific locales."[14]

National Security Implications

Because of differences of opinion regarding production, climate, and population trends over the next quarter century, it is difficult to make judgments regarding the national security implications of predicted global food shortages. If the more extreme starvation scenarios prove to be correct, it is not difficult to imagine the outbreak of regional conflicts in which affected populations would attempt to seize the land or water resources of comparatively more fortunate neighbors. Such possibilities tend to exacerbate tensions between the industrialized and the have-not nations and may encourage the monopolization of raw materials and other forms of economic warfare. It has been pointed out that the developing countries contain two-thirds of the world's population but consume only one-fourth of the world's protein, and most of that in the form of cereals. By contrast, the 1 billion people comprising the richest nations use almost as much cereal grain for feeding their livestock as the 2 billion in the poorest countries use for strictly human consumption. The dependence of the poor populations on imported food could be double the 1970 level by 1985.[15]

Considering that key food-deficit areas are also areas of underdevelopment, it is clear that strategically valuable raw materials—where exportable—are the only bargaining lever of such regions. Inattention to this aspect of the global food situation on the part of the United States and other Western industrialized nations could introduce undesirable new complications into world politics. Even outside the scope of adversary relationships, protective agricultural price policies at the national level can disturb the international environment and create misunderstandings. Lester Brown cites several examples of this phenomenon: the temporary U.S. embargo on soybean exports in 1973, implemented as a result of domestic price considerations, provided an immediate irritant to U.S. relations with Japan and Europe; a rice embargo by Thailand, implemented as a result of production shortfalls in order to preserve domestic supplies, "wreaked havoc" in terms of runaway prices elsewhere in Asia; Brazil's recently imposed export restrictions on soybeans and beef will probably cause similar problems.[16] It is a logical prediction that if global shortages become more critical such unilateral moves will have greater impact, and international misunderstandings will become more frequent and intense.

Tactics versus Strategy

The global food situation has so many ramifications that tactical solutions to some of the problems may get confused with long-term strategic requirements. The difficulty of the challenge should not preclude the development of rational strategies. At this relatively early date following the food and energy crises of the early 1970s, considerable attention in the United States is being directed toward certain components of the macro-problem. Research efforts and policy debates are focusing on such diverse areas as climatology, demography, projected resource scarcities, food as an element of U.S.-Soviet relations, new demands on agricultural technology, domestic agricultural policies, and the moral as well as practical implications of international food assistance programs. The inadequacy of these efforts stems from the fact that policymaking at the national level is better attuned to specific, short-term problems. Taken in isolation, agricultural price subsidies and surplus problems can be resolved through congressional-executive branch action; the foreign policy constituency can make and remake rules concerning such political issues as Soviet relations and the political-humanitarian problems of aid to the developing world. Nevertheless, essentially every decision on food policy requires compromise. The marginal utility of competing options may not be too difficult to calculate over the short term. For example, contractual grain sales to the USSR provide double payoffs in both economic and political terms, and the manipulation of domestic subsidies in the United States helps to balance competing demands from farmers and consumers. But what may appear to be an optimum compromise at the moment can be a colossal misstep in the context of future national interests.

Obviously it is difficult to suggest an agenda for researching and eventually developing a logical set of options for a futuristic U.S. food policy. It is possible, however, even at this stage, to define areas in which greater knowledge could contribute to a better understanding of the problem. These areas include better forecasting of the world's climatic trends, the development of more accurate population projections, better analysis of the capabilities and limitations of the green revolution in offsetting predicted food shortages, and an updated assessment of U.S. strengths and weaknesses in the areas of agricultural productivity and raw materials requirements.

The Climatic Variable

Climatology is one science that may help us forecast the relative strength of U.S. agriculture in the coming decades. But certain disagreements among the world's leading climatologists must be at least partly resolved before a reliable prediction can be made. According to one theory, the earth is undergoing a cooling trend that will reverse what has been a period of basically benign climate for at least the last three or four decades. Although our knowledge about the actual effect of such a cooling trend on agricultural production is limited, there is speculation that the reduced growing season in the northern latitudes would necessitate a southward shift in the growing of some crops and the substitution of lower-yield varieties in the affected areas. Concurrently, the slight reduction in temperatures in areas below latitude 45 might result in higher production of such crops as corn and soybeans. Because of its favorable geographic location and large land mass, the United States could reap the advantage of even greater productivity, while Canada and Europe might be adversely affected. Modern agriculture could probably cope with a gradual decrease of one degree centigrade in average annual global temperature through modifications in plant breeding and seed production. However, abrupt or extreme changes in temperature and precipitation could have extremely negative effects on agriculture.[17]

According to one version of the cooling theory, the return of a nineteenth century-type climate would mean that by the 1980s there would be broad belts of excess and deficit rainfall in the middle latitudes, accompanied by more frequent failure of the monsoons in South Asia, China, and western Africa, and shorter growing seasons in Canada, the USSR, and north China. Because of the geographic locations of the major grain-producing countries, only the United States, Australia, and Argentina would escape adverse effects. The USSR, China, and South Asia would need to import large quantities of grain. If the cooling trend proved to be marked and persistent, actual food shortages would result in spite of any remedial price or distribution arrangements. This theory represents a "worst case" in terms of global interests and a case of mixed blessings in terms of the United States, which would have to deal with the sharp and conflicting economic, political, and moral pressures related to the distribution of its food resources. An opposing greenhouse theory holds that the world is actually becoming warmer as a result of the cumulative burning of fossil fuels. According to this theory, the release of massive quantities of carbon dioxide into the atmosphere will lead to an unprecedented noncyclical rise in average temperatures, causing a completely different agricultural situation.

Temperature change per se is not the most serious potential threat to food production. An increase in climatic variability rather than a change in a single direction would actually be a more serious problem. Some observers believe that climatic fluctuations will be greater during the remaining years of this century than they have been during the past three decades. In the view of some experts, these climatic fluctuations could lead to crop failures beyond our ability to control or be mitigated by agricultural technology.[18] One version of the variability theory holds that the U.S. Midwest could be afflicted again with the drought conditions that produced the Dust Bowl in the 1930s. In this event, or possibly under less extreme conditions of change, the United States would be in a less advantageous situation than if a general cooling or warming trend were to occur, but there would still be conflicting domestic and international pressures on its relatively bountiful food resources.

A fourth situation might arise if the foregoing predictions proved to be unfounded and the climatic trend continues within the limits experienced since World War II. In this case climate would not be an important variable in the U.S. agricultural equation. Rather, factors like technology and increases or decreases in foreign production due to nonclimatological reasons would assume greater relative importance.

U.S. policymakers will have to consider the basic question of the degree to which modern agricultural technology can reduce the adverse effects of climate. Some experts believe the potential impact is great; others contend that the large increase in crop yields that has occurred since 1940 " . . . has been achieved as much by a long run of favorable weather as by improvements in plant breeding and increases in fertilizer application."[19]

The Population Variable

Equally important, and perhaps more controversial for planning purposes, is the question of world demographic trends. Many of the dire predictions that have been made regarding future food crises are based on Malthusian-type population figures. These extreme predictions ignore the possibility of counter trends and an eventual slowing of the global birth rate to supportable levels. Focal points of the controversy include such issues as the likely effectiveness of family planning efforts; the time-span over which inevitable declines in the present birth rate will occur; the effects of natural, though harsh, checks on growth, such as increasingly widespread malnutrition, hunger, and disease; and the relationship between regional birth rates and economic or food resources. In addition, the pressure on food resources from actual population growth on the one hand must be distinguished from the pressure of affluence within lower income strata on the other.

An Oak Ridge National Laboratory study theorized that the world population could never exceed 15 billion; the shortage of at least one irreplaceable resource would render any further increase impossible. The study points out, however, that the widening gap between rich and poor nations would, in the meantime, lead to mass starvation and war, thus checking population growth long before the 15 billion population level could be reached.[20] A less pessimistic note has been sounded by certain participants in the Club of Rome study who contend that it is impossible to predict with any certainty a country's growth over the long run. They say that, given the uncertainties involved, decisions in governmental or other spheres based on such projections merely represent gambles. The message of most doomsday authors is not that their forecasts are inevitable, but that they could become so if appropriate action is not taken now.[21] Viewing the situation with comparative optimism, such futurists as Herman Kahn believe that these doomsday projections are, in fact, based on an aberration in long-term global population trends: that the world population is still showing the effects (actually temporary) of the industrial revolution.[22] According to this view, the developing countries will follow the path of the advanced nations as their rising affluence eventually dampens high birth rates.

Whether affluence is a factor or not is itself the subject of controversy. Although the long-term impact of affluence may fit the optimists' picture, recent experience in the developing countries suggests that per capita income must increase at an extremely high rate per annum before demands for more and better food taper off. International development assistance programs thus can be viewed as another variable in the already complicated equation involving rising affluence, changing demands for food, and declining birth rates.

Illustrating the impact of rising affluence on food demand, Lester Brown discusses the "income elasticity coefficient of demand," a term generally used by economists to describe the relationship between changes in income and changes in per capita food consumption.[23] This coefficient is a numerical expression of the percentage increase in income that is spent on food. For example, the world grain supply is only growing at an annual average of about 30 million tons, or less than 2.5 percent per year. Of this annual growth in supply roughly 1.9 to 2.0 percent, or 24 million tons, are required to meet the demands of population growth; about 6 million tons (i.e., about one-fifth) are required to meet the demands of rising affluence.[24]

Food: Tradeable Asset in an Interdependent World?

What is the role of food as a tradeable asset in an era of raw materials shortages? It is a perplexing question. In the new context of North-South relations (as opposed to East-West), the labelling of have and have-not nations has become difficult. The industrialized countries of the North have exportable capital, technology, and, in some instances, foodstuffs. The developing countries of the South lack these assets but have exportable raw materials. The OPEC nations obviously have both exportable capital and at least one highly valued raw material.

There is little doubt that the United States' enormous

agricultural productivity constitutes a valuable asset in the changing balance of global wealth, but the problems of utilizing this asset to maximum advantage are more complex than they would at first appear. For example, U.S. agriculture is energy-intensive; assuming that alternate energy sources will be available in quantity only in the distant future, the costs of producing and transporting food will change significantly in the event of further oil price increases, especially if they are accompanied by boycotts. The export earnings of U.S. agriculture will be overshadowed by the rising costs of petroleum imports.

Food is not directly tradeable for oil since the food import needs of the petroleum producing nations are in fact minimal when compared to the demand for their principal export item. Likewise, international food markets do not lend themselves to monopolization in the same manner as the petroleum industry. Both commodities constitute rather clumsy weapons in terms of the damage that their usage can cause to the overall world economy, but oil is easier to manipulate. The issues would become more charged if the developing countries attempted to form cartels for other key raw material exports. Through substitution, stockpiling, and other efforts on the part of the industrialized nations, such cartels would probably prove less effective than OPEC's. It would not be surprising to find food used as a "counter-weapon." The concept of food as a weapon contrasts starkly, of course, with the idea that it can play a central, more positive role in international development efforts aimed at eliminating the very problems that could trigger economic warfare.

2
The Climatic Variable

Attempting to construct a plausible model of world climate to be incorporated in a global food policy is perhaps more difficult than trying to build a model for population growth. Whereas the latter involves primarily statistical interpretation and a certain amount of guesswork as to the timing of events that will alter birth rate patterns, long-range climate prediction is a less exact science that suffers from several conflicting theories. There is disagreement among respected world authorities on the scientific level, and there are disputes over the degree to which climate has retained its importance as a key variable in world agriculture. For example, it has been argued that climate poses a smaller threat to future production than it once did because of the advancement of agricultural technology, the increased accuracy of forecasting methods, and new possibilities for weather modification[1] –at least over the short term.[2]

Considering the importance of the question of whether the world will be able to feed itself during the balance of the twentieth century, it seems unnecessarily risky to dismiss the importance of climate with these three arguments. It is preferable to concentrate on the more important issue of what meteorological trends will actually be experienced between now and the year 2000. This is a critical matter for U.S. domestic agriculture and the world grain trade.

Climate and U.S. Agricultural Strength

Agriculture and climate experts already have a fair knowledge of the general climatic factors affecting crop yields in the mid-latitude regions of the world, where large quantities of grain are produced. The Bellagio Conference Report summarized these general factors and noted some of their specific effects on certain U.S. crops:

> Much of the world's grain is produced in mid-latitude regions where summer temperatures average between 70° and 75°F (21° and 24°C). The geographical upper limit of these regions is dictated by the length of the growing season and the lower limit by high summer temperatures. Within these regions there is no doubt that weather conditions affect the crop yields in particular areas from year to year. Generally speaking, the highest yields occur when the summer temperatures are lower than normal. Higher rainfall usually accompanies cooler summer weather. In many years, these climatic influences tend to balance out geographically, with dry, hot weather in one growing area and cooler, wetter conditions in another.
>
> The yield of wheat grown anywhere in the United States is reduced by higher than normal temperatures from flowering time until the crop is mature. Great Plains wheat produces higher yields in response to greater than normal rainfall, but wheat grown in the corn-belt states (such as Illinois and Indiana) gives greater yields when the rainfall is normal or slightly below.[3]

According to one hypothesis, the impressive improvements in U.S. agricultural yields since the early 1950s are the result of unusually favorable and stable climatic conditions, rather than any technology advances.[4] If climate is indeed this important to U.S. agricultural strength, it is all the more essential to consider what changes, if any, will occur over a comparable period in the future. Careful assessments will have to be made involving difficult choices among hypotheses that depict the worldwide climate as cooling, warming, or subject to greater variability from a variety of causes.

The hypothesis that the world has just emerged from an abnormally favorable period could have special significance for the Soviet Union and, indirectly, for the United States because of its present grain sale commitments. In a dialogue that took place at the Kennan Institute for Advanced Russian Studies at the Woodrow Wilson International Center for Scholars on 16 November 1976, U.S. experts contended that as a result of adverse climatic changes the USSR might fail by a significant margin to meet its agricultural output goals by 1980. The Soviets' 1976-80 agricultural plan calls for an increase in production, bringing the totals to 215-20 million metric tons annually. The 1971-75 average was 181.5. According to this estimate, target shortfalls could range from 10 to 20 million metric tons per annum, irregardless of the 1976 bumper crop of 224 million metric tons. Such a shortfall would prompt the USSR to make grain purchases of 15 to 20 million metric tons a year from the United States over the next five years. It was also noted at the conference that during 1971-75 the USSR had three good crops and one average or slightly below average, in addition to the 1975 crop of only 131.9 million metric tons. A more reasonable expectation for a five-year period would be two good crops, one average, and two poor. One U.S. expert observed that favorable climatic conditions accounted for about one-half of the Soviet production increase between 1962 and 1973. Although rainfall in the so-called virgin lands of Kazahkstan was only 4 percent below long-term averages, the region was considered to be undergoing a major drought because its rainfall was compared with levels in recent years. A senior Soviet participant at the symposium disputed the Americans' forecasts, claiming that increasing productivity would enable the USSR to meet its 1976-80 planning goals. The implications of the disagreement are, of course, more than academic. If the future production of the USSR falls within the more pessimistic parameters forecast by the American side, it could push the Soviet demand for U.S. grain to the predicted 15-20 million metric ton level, with significant impact on U.S.

domestic prices and the remainder of our export trade.[5] Indeed, the more pessimistic prediction for 1977 proved correct. The 194 million ton Soviet harvest fell a full 20 million tons below estimates. This problem is a fine illustration of the roles of climatic variability and long-range forecasting in the global food equation.

The Cooling Theory

Perhaps the most dramatic and well-publicized climatic change hypothesis relates to a suspected cooling trend. Several prominent climatologists agree that such a trend may be occurring in the Northern Hemisphere and could lead to a decrease of two or three degrees Fahrenheit in average hemispheric temperature by the end of the century. Although this amount might seem insignificant to the layman, it could have an extraordinary impact on U.S. and world agriculture. Proponents of the cooling hypothesis base their predictions on several indicators, including the fact that temperatures in Iceland—a pivotal measuring location—have been abnormally warmer in the past four decades; warmer than at any time in the past 1,000 years. Other evidence includes a perceptible shortening in the English growing season since the 1940s and recent increases in the Arctic ice areas, including glacial expansion.

The carbon dating of fossil soil in northern Canada has provided new insight into the history of cooling cycles. Major climatic shifts have occurred more than a dozen times during the past 1,600 years. During this period, maximum temperature drops tended to occur within 40 years of the beginning of each downturn; the earliest return to normal required 70 years. If these facts are significant, global long-term climate patterns resemble those of the nineteenth century more closely than they do the patterns of recent decades. Assuming that a downturn began in the 1960s, one could expect to find, within two decades, broad belts of excess and deficit precipitation in the middle latitudes; increasing failures in the monsoons that affect agriculture in the Indian subcontinent, southern China, and western Africa; and shorter growing seasons in Canada, northern Russia, and northern China. One could expect that Europe would be colder and wetter and only the United States and Argentina would remain relatively unaffected.[6]

According to a leading proponent of the cooling theory, a long-term cooling trend began around A.D. 1600 and continued until the beginning of this century. It was interrupted by only one temporary warming period that ended about 1945. Reid Bryson points out that since 1945, the average temperature of the Northern Hemisphere has declined nearly as much as it had previously risen; the average temperature of Iceland has declined to its former level. Since 1951, the temperature of the whole North Atlantic has declined about one-eighth of the difference between recent and full glacial temperatures, with a resultant southward shifting of the Gulf Stream; the English growing season has diminished by two weeks; droughts in north-west India have become more frequent; monsoons have gradual-ly retreated toward the equator in west Africa (culminating in seven years of famine); midsummer frosts have returned to the U.S. upper Midwest; severe icing has occurred in the Canadian Arctic; the snow and ice cover of the entire Northern Hemisphere increased by about 13 percent in the winter of 1971–72 and has remained at that level; and finally, offsetting changes have been noted in the global wind systems (circumpolar vortex and the subtropical anticyclones).[7]

Although the science of climatology cannot yet offer totally adequate predictive models, this and other evidence of a cooling trend could be taken into consideration as a basic scenario during multi-decade planning. If the predictions are correct, there are several serious implications for U.S. agriculture. U.S. production might be unaffected or perhaps enhanced, whereas that of Canada and the Soviet Union would be adversely affected by a shorter growing season. The production goals of the Soviet Union, currently centered on providing an adequate feed-grain supply for an expanded livestock industry and on maintaining human consumption requirements, would be even more difficult to meet given domestic shortfalls and reduced export offerings from a similarly affected Canada. Concurrently, the lack of monsoons in South and Southeast Asia would reduce grain output there. In the case of South Asia an already burdensome foreign aid problem would be compounded. Both northern and southern China would be adversely affected, adding to global demand. Although resultant high world grain prices presumably would favor U.S. commercial export trade, there would be simultaneous appeals for increased U.S. food aid. Many developing countries, due to their tropical locations, would face more frequent droughts and most likely would fail in their efforts to expand agricultural output to keep pace with their population growth.

Many major dams and irrigation systems built during the recent favorable decades are dependent on prevailing rainfall patterns. Most of the hybrids and all of the green revolution plant strains were developed to take advantage of the increased warmth and moisture that have been common during this period. The overall magnitude of a climate reversal is, of course, impossible to predict. A critical question relates to the suddenness and extent of the cooling process, since adaptive plant breeding and unconventional food sources could be developed only on a gradual basis.[8] Proponents of the cooling hypothesis point out that climatic changes tend to be rapid rather than gradual and that cool periods in the earth's history have been periods of greater than normal climatic instability.[9]

The Greenhouse Theory

The extent of the controversy among leading climatologists is underscored by the global greenhouse hypothesis, which predicts a long-term warming rather than a cooling trend. The greenhouse theory is based on the assumption that the earth's atmosphere has a very fragile carbon dioxide (CO_2) balance that may be undergoing significant changes as a result of man's use of fossil fuels. Stated simply, the sunlight reaching earth allows about 98 percent of the planet's water to remain in a liquid state; the balance exists either as solid ice or as atmospheric gas (a miniscule .001 percent). The normal level of atmospheric carbon dioxide is only 330 parts per million, but vast additional quantities are stored in the oceans and in limestone rock. The CO_2 balance is influenced by green plants, which consume gas in their growth process and release it during the subsequent decaying cycle. In the long-term processes of creating such combustibles as food and fossil fuels, CO_2 is concentrated and prevented from returning to the atmosphere. Coal and petroleum represent a massive storage of carbon dioxide, which was captured millions of years ago when the earth had a predominantly tropical climate. The tropical climate, in turn, resulted from extensive volcanic activity and the release of large quantities of carbon dioxide. As a result of the cumulative release of massive amounts of stored carbon dioxide from the burning of fossil fuels in the industrial and current post-industrial era, some theorists fear that a carbon dioxide buildup in the atmosphere will lead to a warming trend with serious consequences. For example, the rise in fossil energy consumption of about 5 percent per year is estimated to have increased the atmospheric CO_2 content from 290 parts per million in 1900 to almost 330 parts per million today; the level could reach 379 parts per million by the year 2000. If that were to happen, average global temperatures would rise by one-half degree centigrade and by two degrees within a century.

Some scientists foresee other complications, including a runaway greenhouse phenomenon as the warming oceans release additional carbon dioxide, as pollutants accumulate in the upper atmosphere and break down into new chemical compounds with unknown effects, and as expanding urbanization turns cities into "heat islands." There are also conflicting theories as to whether the release of particulate matter into the atmosphere will have a warming or cooling influence on the climate.

The basic greenhouse premise has not gone unchallenged; its critics say that temperatures do not increase in proportion to an atmospheric increase in CO_2 and that the consumption of heat-producing energy may not be increasing at the 5 percent rate the pessimists have predicted.[10] It has been argued that the total production and use of energy by humans is only slightly more than .01 of 1 percent of the sun's heat absorbed at the earth's surface and that no significant global climatic warming has occurred as a direct result of mankind's release of heat. However, others argue that this is not a uniform process. Heat released to the environment by the urban megalopolis stretching from Boston to Washington, D.C., for example, has grown to an average of 5 percent of the net energy input by natural radiation. In the view of some scientists, a regional heat island already has been created.[11]

The Variability Theory

Perhaps the hypothesis of greater variability in global climate should be of more concern in food policy formation than either the cooling or warming predictions. Whereas one consistent trend or the other could be at least partially counteracted by adaptive plant breeding, by the shifting of cultivation to more suitable climate zones, or by other agricultural strategies, variability increases the problem of prediction and tends to negate efforts to adapt plants through selective breeding. At the recent Bellagio Conference on the effects of climatic change on food production and interstate conflict, several experts observed that although the earth's climate has always been changeable, there is no single theory—or even a combination of theories—that completely explains the fluctuations and changes. They agreed, however, that important variability factors include periodic infusions of volcanic dust into the atmosphere, changes in ocean surface temperatures, alterations in the Arctic and Antarctic ice masses, negative and positive "feedback loops" involving winds and ocean currents, and human and extraterrestrial inputs into the earth's atmosphere. Because of these factors, some atmospheric scientists have maintained that "the climate system is inherently unstable and its behavior is *unpredictable* [italics added] because its fluctuations are random. However, it appears unlikely that the system is inherently random."[12] The key problem for climatologists and agriculturists alike is to discern what patterns exist.

One offshoot of the variability hypothesis relates to sunspot cycles. Research efforts in England suggest a positive correlation between the occurrence of sunspots and increased agricultural productivity, at least for the Northern Hemisphere as observed over the past two or three decades. John Gribbin directs further speculation to the "double sunspot cycle" of approximately twenty-two years based on evidence of drought occurring in some areas of the United States in 1954 and again in 1974, when sunspots were minimal. There may be 90- and 180-year periods to the solar cycle. If this hypothesis is correct and other factors do not intervene, the global climate can be expected to deteriorate between the early 1980s and the beginning of the next century. This would involve, among other things, a continuing southward drift of rainfall belts in Africa and a continuation of the drought which had such a dramatic impact in the Sahel region. If the sunspot hypothesis is valid, Gribbin notes the predicted adverse climatic effects for much of the remainder of the century could be worsened by a concurrent cooling trend.[13]

Whatever the cause, increased climatic variability could result in a U.S. drought reminiscent of that in the 1930s. This would be of considerably greater global significance than the recent African drought in view of the international importance of U.S. agriculture. There is some disagreement about whether the effects would be as adverse as those experienced in the Dust Bowl era because of improved soil management practices and water conservation techniques. There is less doubt that drought conditions would result in reduced production and that this reduction could be serious because of simultaneous crop shortfalls in other parts of the world and a growing global dependence on U.S. productivity.[14] Although extended drought would cut into the traditional U.S. agricultural advantage only in a marginal sense, it is a spectre that serves as a reminder that even vast agricultural resources cannot give a nation immunity from climatic change.

Technology to the Rescue?

With the inevitable confusion resulting from the conflicting cooling, warming, and variability theories, it is logical to ask whether technology—in the form of weather modification techniques—can provide mankind with the ability to mitigate the worst effects of climatic change. Unfortunately, there is a substantial diversity of opinion among scientists on the present state of the art for specific types of weather modification and for weather modification in general. Despite initial modest successes in this field, which began in 1946, scientific knowledge of atmospheric processes remains rudimentary, and present cloud-seeding techniques do not permit weather control "on demand."[15] The results have varied depending on the types of cloud systems, geographic locations, and whether the efforts were aimed at creating rainfall or mitigating the effects of storms.

Most of the other modification techniques that appear physically feasible have not been explored enough to evaluate their impact. For example, cloud clearance operations appear to be practical on a limited scale and might prove useful for fruit and vegetable production where short periods of additional sunlight are beneficial. But there is evidence that the effects of some weather modification experiments may extend over considerably wider geographic areas than those for which they were intended.[16] These problems are considerable on the national scale and become formidable in the international environment. There is a very real risk that modification programs undertaken in one country will be blamed for causing unusual, adverse weather in another country regardless of whether there is a scientific basis for such charges. Programs cannot be evaluated merely on economic and scientific grounds, then, but must be considered in a political context as well. Since the world food problem requires international understanding and cooperation, any perceived advantages from weather modification efforts must be weighed in relation to their possible political impact and potentially complicating effects on foreign relations.

A discussion of the relationship between climatic change and food production would be incomplete without mention of an on-going controversy over forecasting methods. The U.S. Department of Agriculture's (USDA's) monthly, seasonal, and annual forecasts of crop yields for the United States and foreign countries are the most thoroughly researched and widely respected predictions on future food supplies in the world. However, some climatologists have criticized the USDA's statistical reliance on past yields and the methodological assumption that future weather will be normal because they say these approaches sometimes result in overestimates of crop yields. Additionally, they feel that application of the normal weather assumption to other data relating to technological inputs could lead to erroneous conclusions about the role of technology and overlook the possibility that unusually favorable and reliable climatic conditions have played a hidden but important role in increasing U.S. agricultural productivity since the early 1950s.[17] These criticisms are more or less futile in the absence of demonstrably better predictive methods.

Linking Research to Policy

Since climate remains such a critical but little-understood variable in food policy decisionmaking, there is a manifest need for better-coordinated investigations and the resolution of conflicting theories in the field. This need has been recognized in the United States and abroad, but progress to date has been limited. One of the most recent and promising efforts to resolve theoretical conflicts has been mounted by the National Defense University (NDU) in Washington, D.C. The unclassified study, funded by the Advanced Research Projects Agency of the

Department of Defense, is based on recognition that climatic changes over the next quarter century could have adverse implications for U.S. policies and programs relating to food production and reserves, for agricultural technology, aid programs, and many other areas. Such changes could also significantly, though not necessarily adversely, affect U.S. national security and international relations, as well as strategic planning. The study generally focuses on identifying the potential effects of climatological variables on world agricultural production in order to produce probability statements evaluating the risks in relation to the offsetting costs. Specifically, the study is attempting to define effects of climate on U.S. and world food production to the year 2000; to generate and evaluate a range of policy options for coping with the consequences of climatic variability; and to identify the particular variables that are of special importance in the policy process.

In pursuit of this rather ambitious goal, the NDU research team has elicited the cooperation of twenty-eight leading climatologists in the United States and abroad who have jointly addressed—through computer conferencing and other interfacing techniques—a set of standardized questions designed to reduce or eliminate some of the disagreements currently plaguing the discipline. On the basis of preliminary results, the NDU researchers plan to generate four to six global scenarios. Each scenario should help define a range of national policy issues (such as levels of food reserves, agricultural land use, foreign aid, economic interdependencies, political alliances, and military strategies) that might confront policymakers during the remainder of this century. The final report, like the project itself, will be unclassified and available to all individuals and institutions concerned with the interrelated problems of climatology and world food production.

Other research is also under way. For example, the National Academy of Sciences has published a world food and nutrition study that covers the climatological issues in considerable detail.[18] Intensive investigations of the competing climatic hypotheses are also being undertaken by universities in the

United States and abroad and by such governmental entities as the U.S. National Oceanic and Atmospheric Administration (NOAA). The present problem is obviously not one of inattention but rather one of coordinating results and assuring that the results receive an appropriate hearing at the national policy level. This is a formidable challenge. But it will be impossible to devise rational national or global food policies until we have a better understanding of the interplay between climate and food production.

The growing awareness of the relationship between climatic change and the world food problem and its national implications is evidenced by the U.S. Climate Program and the related U.S. Climate Program Act of 1977. The former is a compendium of recommendations issued in December 1974 by the Subcommittee on Climate Change of the Environmental Resources Committee of the Domestic Council, which urged development of a long-term climate program on a global scale. The latter is a bill currently under consideration by Congress. A cooperative legislative-executive effort, the bill has four main objectives. The first is to establish a climatic-impact warning system that would be of direct assistance to government decisionmakers. This would overcome perceived gaps in the present information system now employed by NOAA. The second objective is to improve existing climate prediction beyond the usual single-season range. This will be accomplished primarily by improving statistical techniques that focus on the relationships existing among climatic change and food production, water resources, and energy consumption. The third objective is a long-term commitment to the utilization of computer simulation techniques and mathematical models for governmental planning purposes. And finally, the act would encourage development of a global climatic monitoring system, expanding a smaller-scale system developed and operated by NOAA.[19] The National Climate Program Act, if passed, would help to refute the age-old maxim that everyone talks about the weather but nobody ever does anything about it. Considering the scope of the world's population and food problems, this step could not be more timely.

3
Population and Food

Population may be ranked with climate as one of the key variables affecting the world food balance. Fortunately for long-term planning purposes, the science of demography is not burdened with the fundamental theoretical arguments currently afflicting climatology. In demography, extensive historical data and continuing census results exist as a basis for statistical projections. There is also a fair understanding of the various forces that can be expected to perpetuate or alter present growth rates.

Likewise, there is general agreement that the world's population increased at a very gradual rate over the course of dozens of centuries and that a population "explosion" occurred as a result of the economic impact of the Industrial Revolution. But there is some controversy over the likely duration of this trend and the timing and manner in which it will inevitably have to subside. Reduced to graphic form, growth appears as a flat, nearly horizontal line representing hundreds of years and then swings abruptly upward (representing the 1800s). Viewed only in the time-frame of the last two centuries, this upward swing suggests the geometric growth rates and attendant disasters predicted by Malthus. Indeed, this recent upward swing, projected against new assessments of the earth's limited resources, has given rise to neo-Malthusian pessimism concerning man's prospects for feeding himself during the coming decades. Some of the more optimistic futurists perceive this growth curve more as an aberrant blip on an otherwise relatively straight line and predict future population stabilization, although at a higher level.

Unfortunately, both the pessimists and the optimists have overstated their respective positions. Uncritical acceptance of one view or the other prompts the world's policymakers to do either too much or too little in response to the problem. The pessimists point out that the present population explosion is unprecedented in world history and that it will prove disastrous for mankind if unchecked by natural or human forces. For their part, the optimists may be correct in assuming a trend toward stabilization at a higher but acceptable level within the foreseeable future, but they appear to be underrating the enormous problems that the world leadership will have to face during the interim fifty or more years. Although these decades may be negligible on a time-scale of many centuries, knowing this does not make the problems that arise during this time any easier to solve. At least there is almost universal agreement that the world food balance is a delicate one and that the present population explosion could tip the scales toward the side of hunger.

Weighing the Facts and Figures

The dimensions of the problem may be illustrated statistically by looking at the 1.1 percent per annum world growth rate in 1930, its rise to 1.7 percent in 1970, and the present rate of about 2 percent. According to the United Nations' "medium forecast," the earth's population today is almost 4 billion and will rise to 6.25 billion by the year 2000. The total could go as high as 7.8 billion if the more pessimistic forecasts are correct. (Admittedly, an end-of-the-century increase to a total of less than 6 billion is also possible and would provide mankind with a welcome but unexpected reprieve from some of the food problems we have discussed.) Whatever the final figure, world population growth currently results more from a declining death rate—primarily in the less developed countries—than from any widespread increase in the birth rate. It is fairly safe to predict a rising demand for food, in the amount of 2.3 percent per annum over the next decade, and comparably large increases in succeeding decades, but more difficult to forecast the degree to which world agriculture will be able to keep pace. The United Nations' medium variant projection for the period of 1950–2000 and explanatory demographic indicators for the same period appear in Tables 1 and 2.

The population problem should be perceived in terms of two separate spheres—the less developed countries versus the more developed ones—rather than in terms of a global average. The declining mortality rate in the less developed countries has produced a startlingly different pattern from that in the more developed countries. Furthermore, today's differences between the two spheres will become more pronounced in future years. Three-fourths of the world's population may be found in the developing areas. During the past twenty-five years death rates have been falling toward the low level of the more developed countries, whereas birth rates have remained twice as high. This means that the ratio of young to old is considerably higher in the developing countries and that accordingly, a greater percentage of inhabitants will be entering childbearing age than in the economically more fortunate countries. The Population Council, employing Thomas Frijka's computer model of world population, which takes into account this age differential, has concluded that the total world population will grow to 6.3 billion by the year 2000 even if the reproduction rates in all countries are brought down to replacement (i.e., stabilization) level as early as 1980. If we arrive at this goal by the end of the century, a plausible but by no means certain event, the world population would be 8.2 billion by 2050. More than 90 percent of the over 4 billion people added between now and the year 2050 would reside in the developing countries. According to this hypothesis, the population figure for India would be 1.4 billion, for Brazil, 266 million; for Bangladesh, 240 million;

Table 1. World Population, 1950-2000

(Millions)

Population	1950	1955	1960	1965	1970	1975	1980	1985	1990	1995	2000
World	2,501	2,722	2,986	3,288	3,610	3,967	4,373	4,816	5,279	5,761	6,253
Developed countries	857	915	976	1,036	1,084	1,132	1,181	1,231	1,278	1,320	1,361
Developing countries	1,644	1,808	2,010	2,251	2,526	2,835	3,192	3,585	4,001	4,441	4,893
PRC	558	605	654	710	772	839	908	973	1,031	1,090	1,148
Excluding PRC	1,086	1,203	1,356	1,541	1,754	1,996	2,284	2,612	2,970	3,351	3,745

Source: U.N. medium variant projection. United Nations Working Paper No. 55, May 1975.

Table 2. Selected World Demographic Indicators, 1950–2000

Item	1950–55	1955–60	1960–65	1965–70	1970–75	1975–80	1980–85	1985–90	1990–95	1995–2000
Average annual population growth rate						*Percent*				
World	1.69	1.85	1.93	1.87	1.89	1.95	1.93	1.84	1.75	1.64
Developed countries	1.30	1.29	1.21	.90	.86	.85	.83	.75	.66	.60
Developing countries	1.90	2.13	2.27	2.30	2.31	2.37	2.32	2.20	2.09	1.94
Excluding PRC	2.07	2.42	2.60	2.62	2.62	2.73	2.73	2.60	2.43	2.26
Five-year change in population						*Million*				
World	221	264	302	322	357	406	443	463	482	492
Developed countries	58	61	60	48	48	49	50	47	42	41
Developing countries	164	202	241	275	309	357	393	416	440	452
Crude birth rates					*Number per 1,000 population*					
World	35.6	34.6	33.7	32.1	31.5	31.1	30.1	28.4	26.8	25.1
Developed countries	22.9	21.9	20.5	18.1	17.2	17.4	17.4	16.8	16.0	15.6
Developing countries	42.1	40.9	39.9	38.4	37.5	36.4	34.6	32.3	30.2	27.8
Crude death rates					*Number per 1,000 population*					
World	18.8	16.4	14.7	13.5	12.8	11.9	11.0	10.2	9.5	8.9
Developed countries	10.1	9.3	9.0	9.1	9.2	9.4	9.6	9.8	9.9	9.9
Developing countries	23.3	19.9	17.4	15.5	14.3	12.8	11.5	10.4	9.4	8.6

Source: U.N. medium variant projection. United Nations Working Paper No. 55, May 1975.

and for Nigeria, 198 million. The figure of 8 billion would appear to be the minimum stabilization level, but this leveling phenomenon might not occur before a figure of 10 to 15 billion were reached.

Although these statistics are alarming in some respects, the eventual stabilization of population at some higher level than the present figure presents quite a different scenario from that posed by the more extreme neo-Malthusians, who predict that world population will climb to astronomical levels on the present curve. The current climb is a result, as has been noted, of decreased mortality rates rather than increased fertility. One must remember, though, that the mortality rate decrease is a one-time shift during which growth tapers off as life spans approach a biological ceiling.[1] More pertinent in terms of the problems that will confront the leaders of the Western industrialized world in the intervening decades is the fact that their nations are reaching this "biological ceiling." The more populous nations of the developing world are not. The effects of medicine on disease are more predictable than the effects of population on the food balance. Accordingly, both the haves and the have-nots must ponder the impact of population growth on food supply in the world community as a whole.

The controversy over the time-frame for stabilization can be illustrated further by looking at the optimistic interpretations of Herman Kahn. He provides high, medium, and low projections of 1.7, 1.3, and 1 percent, respectively, for the world population growth rate by the year 2010. These figures are somewhat higher than the United Nations' medium projection. Comparing these rates with today's rate, Kahn concludes that the problem of exponential growth appears to be solving itself. He adds, "There can, of course, be no certainty that these projections will prove to be accurate, for we have only the available data and demographic theory on which to rely. But these constitute a historical basis for forecasts and strongly suggest that fears of a population explosion should disappear within the next half-century."[2] Kahn also points to the phenomenon of "demographic transition," which he describes as historical experience rather than as theory. Both birth and death rates have tended to level off in the industrial nations, increasingly so as some have reached the super- and post-industrial stages. He agrees with those demographers and economists who believe that the transition will occur, or is even now occurring, in the developing countries on an increasingly foreshortened time-scale. For example, the time required for transition to very low growth rates in Western Europe and North America was 150 years, versus 40 years for the USSR and 25 years for Japan. Pointing to evidence of decline in the crude birth rates in the 1960s in fifteen developing countries and the probable decline in eight more, Kahn suggests that the 60s may have marked the beginning of a decline in the worldwide fertility rate. By this reckoning, the Population Council's "high" prediction of a stabilized world total of 15 billion by the mid-twenty-first century would not occur for approximately another century.[3]

Considering that competing projections vary by several billions and time-frames by as much as a century, it is clear that demography involves guesswork in exponential amounts as one proceeds into the future. Comparing Kahn's perceptions with those of the Population Council, there appears to be a weighting of evidence. Kahn believes that the demographic transition may already be under way in the developing nations, and the Population Council places emphasis on the reproductive potential of these nations due to the predominance of a younger population. This is only one example of differing interpretations that lead to dissimilar projections. Nevertheless, experts on all sides seem to agree that even observable, long-term trends are subject to cyclical variations. For example, the long-term decline of the fertility rate in the United States became sharper in the Depression years and reversed itself in the post-World War II "baby boom." One observer has stated that large fluctuations in fertility, and in mortality as well, are not inconsistent with the long period of near-zero growth that characterizes most of the history of world population.

Although the arithmetic of growth leaves no room for a rate of increase very different from zero in the long-run, the short-term variations in the world population growth rate have been frequent and of considerable extent. World population, which from one perspective appears to have been almost static for thousands of years, actually experienced brief periods of rapid growth, during which it expanded severalfold and then suffered catastrophic setbacks. Since these ups and downs are linked to political, economic, and social variables that are impossible to foresee, even an accurate guess with regard to when and at what level the world population will reach a new stabilization point can be of little consolation to the policymakers who must adapt agriculture to annual demand. In this sense, both the optimists and the pessimists are engaged in an argument that overlooks a serious interim problem for world agriculture. Statistics fail to illustrate the human dimensions of the population-and-food challenge. According to one expert,

> Arithmetic makes a return to a growth rate near zero inevitable before many generations have passed. What is uncertain is not that the future rate of growth will be about zero, but how large the future population will be and *what combination of fertility and mortality will sustain it* [italics added]. The possibilities range from more than 8 children per woman and a life that lasts an average of 15 years, to slightly more than 2 children per woman and a life span that surpasses 75 years.[4]

These are extreme scenarios, but it is a valid point that a biological equilibrium does not necessarily constitute the optimum in terms of mankind's political, economic, or social interests.

The Two Population Spheres

The phenomenon of two population spheres—the underdeveloped and the developed worlds—requires further examination from the perspectives of distribution statistics, age factors, and urban-rural growth patterns. To emphasize further the contrasts already noted, during the past twenty-five years the population increase in the developing countries amounted to five times the increase in the industrialized world. The current annual increase in the less developed countries is more than 60 million inhabitants. Asia, with a land mass smaller than Africa, has 55 percent of the world's population and is home to 75 percent of the 2.9 billion people classified as *developing*. By contrast, Latin America has only 11 percent and Africa, 14 percent of the developing world's population. Growth rates in these areas are continuing at about the same 2.4 to 2.7 percent level, with the exception of China, which may be down to 1.7 percent per annum.[5]

Statistical forecasts for the developed world, which enjoys far fewer problems in terms of population pressures, suggest that growth is a factor to be reckoned with here as well. According to one study:

> Even if replacement fertility for the developed world as a whole were maintained for the foreseeable future beginning in the 1980's, the population, because of the age composition resulting from its past history of growth, would increase by more than a fourth, adding some 300 million more people. Of that total about two-thirds would have been added by the year 2000. . . . What does seem plausible today is that the population of the developed world will average replacement fertility by the early 1980's. If that level is continued in the future, the developed population will continue to increase for a good part of the next century, stabilizing at just below 1.5 billion.[6]

On the one hand, population expansion inevitably will contribute to further increases in both the less- and more-developed countries until their respective stabilization points are reached. On the other hand, cultural differences that influence the childbearing age of women will reinforce the present contrast in growth rates. Whereas twenty to thirty years of age is the normal age span in industrialized societies for childbearing, it begins earlier and lasts longer in pre- and proto-industrial economies. Childbearing age in this sense is defined by economic and social circumstances, but nutrition appears to be playing a role in determining, physiologically, the onset of menstruation—menarche. The Harvard Center of Population Studies has developed evidence of a linkage between quality and quantity of food intake and the achievement of critical body weight. Significantly, the age of menarche has been dropping in most areas during the past century. This could offer an additional complication for countries where the diet is improving but social and economic forces are not yet operating to discourage childbearing at an early age.[7]

The Rural-Urban Dichotomy

The "numerical explosion" has received considerable publicity in recent years, but vital problems relating to the geographical distribution of populations have received comparatively little public and governmental attention. In the Western world, the process of industrialization spurred urbanization at an early date. Mechanization not only provided jobs in cities and towns but also increased agricultural productivity, enabling fewer farmers to feed more city dwellers. By contrast, development in the post-World War II era has entailed migrations from farm to town not because of available industrial jobs but as a result of popular expectations of a more comfortable lifestyle. The new urban inhabitants are politically articulate and—in one fashion or another—receive governmental attention and economic concessions. There is a resultant overemphasis on urban employment problems and a concurrent neglect of agricultural issues. Observers have noted that people with rising expectations prefer almost anything to farming. Thus, in the words of one food scholar, "The quandary faced by most countries of the Third World can be viewed more instructively not as a race between food and population, but as one between population and employment on the one hand and employment and food on the other."[8]

To date, this more sophisticated view of the relationships among population, employment, rural-urban migration, and food appears to have been overlooked, or at least underrated, by policy planners on both the donor and the recipient sides of international development aid. The donors tend to pursue impractical industrialization schemes and other urban-oriented prestige projects at the expense of domestic agricultural development. The recipients tend to support such projects either for political purposes or through ignorance of the consequences in terms of the international food balance. In extreme situations, the populations of food-deficit nations have actually congregated around port cities in order to have better access to foreign food shipments. Despite the horrendous indirect economic costs in the long term, this phenomenon has not been totally unwelcome to officials faced with more immediate and visible hinterland distribution problems. The demographic result has been the creation of densely populated cities with rural cultures. It is apparent that the improvement of agricultural productivity in the Third World, which will be so important to the feeding of future billions, may be impeded by unwise geographic shifts among these populations.

Melding Food and Population Statistics

Since one of the most obvious sources of concern arising from the population explosion is that of global food shortages, it is useful to consider where deficits are likely to occur within a reasonably predictable time span. The European countries, the United States, Canada, Australia, and New Zealand comprise most of the developed world and enjoy either a high agricultural productivity or the income to import food. Thus, the poorer countries appear as the likely problem areas, and their high population growth rate looms as the complicating factor in the task of forecasting deficits. The International Food Policy Research Institute of Washington, D.C. (IFPRI), has taken an imaginative approach by categorizing countries according to their food prospects and creating high and low income scenarios. IFPRI's projections are shown in Table 3.

Table 3. Food Gaps in Developing Market Economies, 1975 and Projected 1990

(Millions of Metric Tons)

Developing Market Economies	1975		1990 Projected Surplus/Deficit			
	Actual[a] Surplus/ Deficit	Additional[b] for 110% of Calorie Standard	At 1975[c] Per Capita Consumption	Low[d] Income Growth	High[e] Income Growth	110% of[b] Calorie Standard
Total – Asia	– 9.3	36.2	– 7.5	–38.4	–49.9	–62.3
Low Income Food Deficit	– 5.5	33.8	–12.5	–35.2	–43.8	–64.6
High Income Food Deficit	– 6.9		– 9.4	–14.4	–16.5	– 8.0
Exporters	+ 3.0	3.0	+14.4	+11.2	+10.4	+10.3
Total-North Africa and Middle East	– 9.6	6.3	–19.5	–29.8	–33.8	–31.4
Low Income Food Deficit	– 3.8	3.2	– 5.4	– 7.0	– 7.5	–11.7
Middle Income Food Deficit	– 1.1	0.2	– 4.6	– 8.6	–10.2	– 5.4
High Income Food Deficit	– 4.7	2.9	– 9.5	–14.2	–16.1	–14.3
Total-Sub-Sahara Africa	– 2.2	12.6	–11.3	–22.3	–27.6	–31.1
Low Income Food Deficit	– 1.8	11.5	–12.3	–22.5	–27.4	–30.6
Middle Income Food Deficit	– .4	1.1	– 0.2	– 0.9	– 1.3	– 1.6
Total-Latin America	0.0	0.9	+19.3	+ 8.3	+ 5.1	+16.7
Low Income Food Deficit	– .3	0.7	– 0.5	– 0.7	– 0.8	– 1.5
Middle Income Food Deficit	– 7.8	0.9	– 0.3	–10.1	–12.9	– 3.4
High Income Food Deficit	– 1.1	0.5	– 2.0	– 2.5	– 2.6	– 2.3
Exporters	+ 9.2	– 1.1	+22.1	+21.6	+21.5	+24.0

Total- Low Income Food Deficit	-11.4	49.2	-30.6	-65.4	-79.5	-108.4
Middle Income Food Deficit	- 9.3	2.2	- 5.1	-19.6	-24.5	-10.4
High Income Food Deficit	-12.7	2.8	-21.0	-31.1	-35.2	-24.7
Total- Developing Market Economies, Food Deficit	-33.4	54.2	-56.7	-116.1	-139.2	-143.5
Total- Exporters	+12.2	+ 2.0	+36.5	+32.9	+31.9	+ 34.3
Total- Developing Market Economies	-21.2	56.2	-20.2	-83.3	-107.3	-109.3

NOTE: These data are condensed from a paper entitled "Food Needs of Developing Countries," which was presented by Nathan M. Koffsky, Interim Director of the International Food Policy Research Institute (IFPRI), at a USDA-sponsored food policy conference at the Pan American Health Organization, Washington, D.C., on 28-29 April 1977. (They will also be shown in a forthcoming research report of IFPRI entitled "Food Needs of Developing Countries: Projections of Production and Consumption to 1990.")

[a]Surpluses and deficits represent the difference between production and consumption.

[b]To raise the diet of the total population of a country to the minimum calorie standard, without reducing the diet of those already above that standard, requires food production which would provide everyone, if equally distributed, 110% of the calorie standard.

[c]The actual 1975 per capita consumption remains constant, while the U.N. medium population growth forecast is applied in order to determine the impact this population growth will have on consumption (demand) in 1990.

[d]This adds to the population impact, the impact a low level of income growth would have on demand for food in 1990.

[e]This adds to the population impact, the impact a high level of income growth would have on demand for food in 1990. These projected demands are based on historical analyses of the relationship between income and food consumption.

Viewed in terms of food requirements, the U.N. medium projection suggests that *if* the populations of the developing countries increase at anticipated rates, their annual import requirements for cereals will rise from 24 million tons in 1970 to about 52 million tons in 1985 and approximately 111 million tons by the year 2000. If, by great efforts, the developing countries also increase their indigenous food production of cereals from 290 million tons in 1970 to 434 million tons in 1985 and 667 million tons in 2000, this would still represent an unbearably heavy annual import burden from a monetary standpoint. The alternatives for the developing countries would be to abandon dietary improvements, accept a significant rise in death rates, increase food production to levels now believed to be impractical, or reduce population growth more rapidly than the U.N. medium projection. With growth rates conforming to the U.N. low projection, cereal import requirements for these countries would be about 37 million tons for 1985 and 31 million tons by the year 2000. Adoption of more drastic— and presumably less likely—measures to attain a two-child family average (the replacement level of fertility) in most less developed countries would lower the potential demand to 30 million tons by 1985 and create the possibility of a small surplus (about 8 million tons) by the year 2000.[9]

The rate of progress in this race between population and increased agricultural productivity is not especially encouraging. The U.N. Food and Agricultural Organization's Indicative World Plan of the early 1960s estimated that total food production in the non-communist less developed countries would have to register an annual increase of 3.9 percent from 1962 to 1965; however, by 1971 the rate moved only to 2.8 percent. It has subsequently remained at about this level. Meanwhile, the demand for food has increased as a result of rising incomes and concurrent urbanization. The Indicative World Plan envisions a solution to the problem only through breakthroughs in cereal productivity, reversal of the growing "protein gap," increased employment in the agricultural sector, and improvements in the less developed countries' agricultural trade. It should be noted that these and other statistical projections can only point to perceived problems and general trends. They cannot provide precise predictions. In almost all developing countries, food production statistics are normally based on the opinion of local government officials and agricultural officers. Reliable statistics are obtainable only where food products are exported or large marketing organizations exist.

The Affluence Factor

The role of affluence is a curious one because it checks population growth on the one hand and increases demand for food on the other. One food expert has argued that as much as three-quarters of the recent increase in global food demand was the result of population growth and the remainder stemmed from the affluence factor. According to the same authority, "Roughly half the increased requirement for grain comes from growing demand in developed countries, where the population increase rate is low but affluent consumption is high, and half from the developing countries, where the opposite is the case."[10] By contrast, Herman Kahn emphasizes the positive function of affluence, which he considers to be an ally, not an enemy, to the cause of restraint.[11] The apparent conflict between the two viewpoints does not mean that one or the other is incorrect but merely that the affluence factor must be considered by policy-makers in two distinct contexts: food requirements and population growth. Its effect on food is immediate and negative, whereas its effect on population growth would come into play over a longer period and could be viewed as positive. One of the primary problems of rising affluence in the immediate future results from a concurrent rise in demand for meat, which, in turn, involves making trade-offs in the production of grains for humans and for animals. The trade-offs can be decided on the basis of the realities of the marketplace, but inevitable future population pressures and a widening gap between the have and the have-not nations will interject new political and possibly humanitarian issues into the problem.

Lester Brown has provided some interesting statistics on the positive linkage between rising affluence and grain consumption. The amount consumed directly increases until per capita income approaches $500 per year, whereupon it begins to decline, eventually leveling off at 150 pounds. Yet the total amount consumed, both directly and indirectly (i.e., through conversion to meat), continues to rise rapidly as per capita income climbs. Brown points out that as yet, no nation appears to have reached a level of affluence where its per capita grain requirements have stopped rising.[12]

Insofar as rising affluence may limit family size, one wonders how soon this will occur and what the relative population sizes and consequent competing demands on food resources are likely to be in the developed and underdeveloped countries. On balance, the positive effects of affluence would appear to be considerably more ephemeral than the negative ones. The authors of *The Limits to Growth* point out that the social and educational changes that ultimately lower the birth rate and are associated with increasing industrialization take a long time to come about. Furthermore, rising industrialization implies better health, so the number of births could (temporarily) increase as GNP increases.[13] It remains to be seen whether rural development abroad will be more successful than it has been in the past in reducing the traditional lag between increased affluence and slower population growth.

Foreign Aid Challenges

The aid policies which the developed, food exporting nations adopt toward the food-deficit nations will represent an important element in the food-population equation during the coming decades. Massive and sustained foreign aid is a comparatively recent historical phenomenon, dating only from the end of World War II. The United States has played a dominant role in providing both direct development and food aid and indirect assistance through international organizations. It is reasonable to assume that the global food, energy, and population challenges of the final quarter of the twentieth century will pose foreign assistance problems quite distinct from those experienced in the years between World War II and the mid-1970s.

One issue that must be considered is the extent to which programs perceived to have short-term political and humanitarian benefits are actually judicious from the long-term standpoint of the world population problem. Additionally, consideration must be given to the scope and effectiveness of family planning and other approaches to the population problem, for these are becoming components of bilateral and multilateral foreign assistance.

Lester Brown has observed that, historically, human fertility does not decline until certain basic social needs are satisfied. Examples of such needs are an assured food supply, reduced infant mortality rates, literacy, and at least rudimentary health services.[14] These are immense problem areas, and it is not surprising that in only a quarter century of international experience it has not been possible to narrow the gap between the developed and developing nations. So far, advanced societies have been successful in the areas of health and urbanization-industrialization and have lagged in the areas of population control and agriculture. For example, family planning services have been available under the aegis of official national policy in the United States only since the passage of the Family Planning and Population Research Act in 1970. Family planning on an international scale poses a challenge that no nation or organization is adequately equipped to handle. The United Nations Fund for Population Activities estimates that 50 cents to 1 dollar per capita per year for an entire population is required to provide adequate family planning services. The present flow of $250 million per year through this organization is only one-fourth of what is needed.[15] In his response to *The Limits to Growth* study, Mihajlo Mesarovic adds that even with the strongest population control measures, by the year 2000 the South Asian population will increase by more than half a billion. Analysis of a whole set of scenarios from most optimistic to most pessimistic shows that the increase in regional agricultural production, barring massive foreign aid for agricultural development, will not be commensurate with Asia's population increase. A deficit in the food supply of that region can therefore be considered a certainty. Mesarovic adds that equilibrium in the South can be reached only seventy-five years after fully effective policies are initiated. The equilibrium level will be more than twice as high as when the programs began. Accordingly, there is a need to look at least fifty years ahead when considering worldwide development. Mesarovic and his colleagues reached these conclusions assuming initiation of effective action by 1975. Alternate scenarios based on programs beginning in 1985 and 1995 suggest that the first ten-year delay would increase the equilibrium population in the South by 1.7 billion, the second ten-year delay would increase it by an additional 2 billion. Considering that the population of the region was "only" 2 billion in 1950, the prospect of totals of either 8 billion or 10 billion resulting from programs initiated in 1985 and 1995, respectively, is alarming.[16]

Balancing Views

Writing in a special issue of *Scientific American* devoted to population problems, Freedman and Berelson pointed to recent positive developments:

> The first is a growing scientific sophistication about the consequences of population growth, as the simpler answers of 10 to 15 years ago give way to more qualified and more complex ones. The second is a broadening of the definition of the problem to include not only economics but also the terrestrial environment, to regard population growth not only as a burden on the development of underdeveloped countries but also as a multiplier of the stresses on the resources and the environment of developed countries.[17]

Mesarovic is concerned with the costs implicit in delays in implementing effective population control programs throughout the developing world. However, Freedman and Berelson have added a slightly encouraging note by observing that:

> Many countries differ in important ways from their European precursors at a similar stage in the transformation, exhibiting a much more rapid decline in mortality, a greater access to advanced technology and an aspiration to quickly reach demonstrated goals that were attained only slowly by the first arrivals. These differences may facilitate a reduction in fertility . . . in some countries.[18]

But the authors also note that reductions in other areas might have to wait out the process of social transformation.

It seems reasonable to conclude that there is no simple formula available to policymakers in either the developed or developing nations since programs in education, industrial and agricultural development, literacy, health, and family planning will not produce, individually, the results the world requires. Yet, the emerging picture of what world population growth will mean in terms of food and other vital resources may be less ominous than that suggested by the pessimists. The United States appears to be the key international actor because of its food resources and unique position in the sphere of bilateral and multilateral foreign aid and international grain sales. U.S. food and population policies, in the domestic and foreign assistance spheres, will inevitably have an impact on the world population problem.

4
U.S. Food in an Interdependent World

The United States will probably continue to enjoy extraordinary advantages from its bountiful agriculture. Even with only modest increases in productivity, there will certainly be sufficient food for domestic needs and exportable surpluses for trade or aid purposes. There is little doubt that the nation's geographic location, production efficiencies, heavy investment in research, and the absence of a domestic population problem will continue as assets. It is essential from a policy standpoint, however, to understand what the U.S.'s food assets represent in both quantitative and qualitative terms. Taking into consideration that production will probably remain in excess of domestic requirements, there are numerous decisions that will have to be made regarding the manner in which U.S. food reaches the outside world. The dangers of either overestimating or underestimating U.S. food as a strategic, economic, or humanitarian asset have been pointed out.

Implicit Policy Problems

Recently, the suggestion has been made in both governmental and journalistic circles that U.S. food constitutes a "weapon" and that it should be used as such on the international economic "battlefield." This suggestion is linked to a legitimate anxiety over increasing U.S. dependence on foreign energy and raw materials, as demonstrated so vividly by the 1972–73 oil embargo and the subsequent shocks to the national economy. The food-as-weapon proposal must be evaluated carefully on several grounds. Would the weapon be effective? Where and when could it be applied? Would its use in one location involve unfavorable trade-offs in others? Would the food weapon be applicable primarily in U.S. relations with the communist world, the OPEC nations, or other developing nations that export raw materials (in a North-South context); against traditional political allies who are commercial rivals, like the Western European nations or Japan; or as an ad hoc weapon to be used in several of the foregoing contexts? How could the weapon concept be reconciled with the historical U.S. treatment of food as a key component of humanitarian assistance programs? The existence of these policy problems lends support to the thesis that it will be impossible to make rational decisions concerning the use of food—whether as a weapon, as a lever, or as a purely economic asset—in the absence of a better picture of what food actually represents.

Policymakers must consider whether or not recent historical experience is a sufficient basis for formulating policy for the final years of the twentieth century. The energy and raw materials trade-offs represent only two of several considerations. Other factors include the limits to which U.S. agriculture can be expanded within finite land and labor constraints and the extent to which various international supply commitments—both in the commercial and the assistance sectors—will allow flexibility.

Food and Energy

In assessing the negotiable value of U.S. food, the most obvious starting point is the energy area. It is actually of double importance, for the use of energy underlies increasing U.S. dependence on foreign oil and constitutes an important element in our agricultural productivity. It is necessary to determine not only whether food is a weapon or lever in the first (adversary) relationship, but also whether vulnerability in one area may extend to the next.

The degree to which U.S. agriculture is energy-intensive is a cause for concern, in terms of economic bargaining and strategic dependency. The casual observer might tend to overlook the fact that, in addition to gas or fuel oil burned on farms for normal heating purposes and the powering of trucks, tractors, and other implements, U.S. agriculture makes significant demands for electricity, petroleum-based fertilizers, and energy-consuming metals that go into fixed structures and movable machinery. The energy required for pumping irrigation water alone is a very large component of total farm usage. With the degree of mechanization now prevailing in U.S. agriculture, 5 to 10 calories of fuel are expended in the production of each food calorie.[1] Actually, this consumption is modest by comparison with that in the remainder of the food cycle. Downstream, prodigious amounts of energy are consumed in the making of food-processing machinery, in the processing itself, in the container industry, and in the manufacture and usage of transport vehicles. Finally, commercial and home refrigeration and cooking constitute essential but easily overlooked elements in the U.S. energy diet. Expressed in kilocalories $\times 10^{12}$, total energy usage on farms increased between 1960 and 1970 from 373.9 to 526.1; in the food-processing sector it increased from 571.5 to 841.9; and in the area of commercial and home food preparation it increased from 1,440.2 to 2,172. The 1970 figures represent 12.8 percent of total energy use in the United States.[2] This figure has climbed to a current level of about 15 percent.

David Pimentel and his colleagues have done considerable research on the energy utilized by U.S. agriculture, focusing on corn, since that crop approximates average caloric inputs and outputs under modern production techniques. (The energy expended in growing it and the caloric value of the crop are neither high nor low when compared with other crops.) He estimated that in 1970, 330 million U.S. acres were planted in crops (excluding cotton and tobacco). Calculated against a total population of over 200 million, this averages about 1.7 acres per person, which must be corrected to 1.4 acres to account for the fact that about 20 percent of U.S. agricultural products is exported. Expressing the energy requirements for farming this land by modern, intensive methods in terms of the equivalent energy of gasoline, energy utilized amounts to 112 gallons per person. This figure rises to 336 gallons when energy requirements for food processing, distribution, and preparation are taken into account. According to Pimentel's calculations, then, if U.S. agricultural methods were used worldwide to feed a population of 4 billion, the annual energy equivalent in gasoline would be 1.464 trillion gallons.[3]

Pimentel also observes that U.S. food production costs are high by world standards when measured in terms of dollar input per kilocalorie of plant product. This, in effect, represents the other side of the energy equation. In the United States the cost of 1,000 kilocalories of plant product is approximately $38, while the equivalent figure for India is about $10. Part of this discrepancy, Pimentel admits, is due to differences in the plant crops grown in the two nations. The principal raw material of U.S. agriculture is fossil fuel, and the labor input is relatively small, about nine hours per crop acre. Accordingly, production costs are quite sensitive to increases in fuel prices. Green revolution agriculture, which has been promoted abroad by the United States and which thus far has offered some hope of offsetting insupportable future demands for U.S. food aid, unfortunately is also energy-intensive, especially in terms of the chemical fertilizers and pesticides used.[4] Energy usage in these areas could be cut by using the green (i.e., natural) manures. In the United States it might be economically advantageous to farm on a smaller scale those lands that are now marginal for highly mechanized methods.

Additional savings could be realized through the following measures: weed control by use of the rotary hoe instead of herbicides; pest control by biologic means rather than insecticides; relaxation of pest control standards to the level of "treat-where-necessary"; changes in cosmetic standards on the part of both retailers and consumers; improved plant breeding for hardiness; production of alternate solar or methane energy; reduced reliance on highly processed foods; and partially reorienting food transport, from trucks to more energy-efficient rail shipment. The treat-where-necessary approach could decrease pesticide use by 35 to 50 percent.[5] There is reason to doubt, however, that these conservation techniques will be employed extensively unless there are further incentives like fuel price increases. Accordingly, it is more realistic to think of U.S. agriculture as having an ongoing vulnerability in terms of its energy requirements. To retain a proper perspective of this vulnerability, it should be noted that agriculture in the developing countries—although considerably less energy-intensive than U.S. agriculture—has been affected even more directly by rising world oil prices. These countries face the more insidious problems of population expansion, obsolete growing methods, and lack of a strong industrial base or other means of generating money to pay their fuel bill. The United States is in an enviable position to the extent that such problems are simply nonexistent here. On the other hand, the existence of a U.S. food surplus and the political need to address the food problems of the developing countries present a burdensome dilemma for policymakers.

Food and OPEC

The Organization of Petroleum Exporting Countries (OPEC) adds an important, new factor to world food production. The future of this cartel and its possible emulation by other raw materials exporting countries will have a bearing on world food developments for years to come. As noted previously, price increases have had the harshest impact in the developing countries, which are least equipped to bear the cost. For them, the problem of price increases will be compounded further if agricultural modernization schemes continue to stress energy-intensive methods. According to the authors of the second Club of Rome report, "The annual excess of revenues to the oil exporting countries will amount to $60 billion, which is about two-thirds of all overseas investment which U.S. firms have acquired up to this time."[6] This is an enormous transfer of wealth, by any standard, and for the United States and many other nations a significantly adverse balance-of-payments.

U.S. food clearly is a welcome means by which to offset this new drain, but the question remains whether a commodity that offsets is an offensive or defensive weapon. The cost of U.S. petroleum imports is now amounting to approximately $45 billion per year and will likely rise considerably within the decade. For the year ending 30 September 1977, agricultural exports amounted to $24 billion; however, imports—primarily processed foods—cost the nation slightly more than $13.7 billion. U.S. food exports to the OPEC nations amounted to only slightly more than $1.5 billion out of the $24 billion export total and therefore represent only a partial offset in a dollar sense. To the extent that these export earnings are more necessary now that the U.S. oil bill has increased, the utilization of U.S. food as either a weapon or a lever has diminished. Certainly there is no viable option of withholding or threatening to withhold U.S. food from the OPEC nations in retaliation for unacceptable price increases, since these nations (except for Indonesia and Nigeria) have relatively small populations that could be fed easily through purchases elsewhere in the international market. Despite suggestions that the United States could pressure other agricultural exporters to take similar actions, it is questionable whether such efforts would be effective or even whether food shipments through secondary and tertiary channels could be monitored, let alone controlled.

The efficacy of food as a weapon in other international contexts should not be judged solely on its lack of impact as a lever in the OPEC countries. One proponent of the strategy of using food as a weapon concedes that there are moral, political, and technical reasons why the deployment of U.S. food-power might backfire, but he maintains that this power is, nonetheless, a reality with "very subtle manifestations." Such manifestations are supposedly demonstrated by the recent Soviet grain purchases and by the extent to which U.S. grain pipelines crisscross the world without regard for geographic or ideologic borders. For example, East Germany has become almost totally dependent on corn from the U.S. Midwest to feed its hog, poultry, and beef populations, and Japan receives more than half of the 20 million tons of grain that it imports annually from the United States. Grain is especially important to developing nations in the tropics, where soil and climate are unfavorable for growing cereal crops. As Dan Morgan states, the United States' agricultural dominance "gives it the power to affect diets, nutrition, and economies around the world—perhaps even to decide who starves and who lives in times of famine and periods of poor harvest. In many countries food imports from the United States have become a major factor in political stability."[7]

Much of the current debate on the existence or utility of U.S. food-power is a search for remedies to the problems that OPEC oil policies and incipient raw materials-scarcities create. Therefore, it is preferable to assess U.S. food and OPEC oil exports in the context of global interdependence. Considering the quasi-adversary relationships that have evolved between OPEC and the industrialized nations—particularly the United States—following the 1973 embargo, OPEC actions during and following the 1974 World Food Conference in Rome suggest that food may provide a useful "neutral" area for inter-bloc cooperation.

Multilateral Alternatives

An OPEC funding offer that surfaced during the 1974 World Food conference sparked creation of the International Fund for Agricultural Development (IFAD), but not without much negotiation. In response to the initial OPEC offering of $500 million subject to a matching contribution from the industrialized nations, the United States pledged $200 million. This commitment, in turn, was subject to a congressionally imposed condition that IFAD subscriptions total at least $1 billion and that OPEC stand by its commitment to subscribe on a fifty-fifty basis. The U.S. director of the World Food Council, John Hannah, was able to obtain the needed pledges from the industrialized nations, including West Germany and Japan, who pledged $50 million each, by early 1976. But on 19 May 1976, OPEC's governing board reduced its commitment to $400 million, thus invalidating the U.S. offer.[8] Following protracted bargaining, the $1 billion target was finally reached in December 1976 with $435.5 million from OPEC, $567 million from the industrialized nations, and $9 million from the Third World.

The IFAD plans to dispense about $350 million per year in loans and grants for Third World agricultural development. It will do this from a small bureaucratic base that will rely on various other U.N. and international development organizations for field support. Replenishment donations will be needed within two years if the fund is to maintain its momentum and make a meaningful contribution in a development area where actual needs and absorptive capacity amount to several billion dollars per year.

Although it is too early to predict whether the IFAD will live up to its subscribers' hopes, it and the World Food Council—which also is a by-product of the 1974 Rome Conference—are timely arrivals on the international scene. The United States and the OPEC nations, the principal exporters of food and petroleum, respectively, now have a forum and an accompanying bureaucratic mechanism to deal with the serious problem of agricultural underdevelopment. Even though the IFAD could prove a disappointment through the cancellation or nonrenewal of needed subscriptions or through political infighting, it has the potential to "internationalize" one aspect of the world food and energy problem. Under the Fund's unique system, the OPEC nations, the industrialized nations, and the Third World recipients each have one-third of the votes to set policies and approve loans. It remains to be seen whether OPEC—and especially the Arab bloc—will pursue a generally conservative "banker's" approach, as has been the case in the World Bank and other international monetary agencies, or will take a more openly pro–Third World stance. The answer will be determined by the bureaucratic evolution of IFAD itself and the relationships that it builds with the parent U.N. bureaucracy and the international banking and development communities. In any event, the importance of IFAD rests in the opportunity it has to tackle global agricultural development problems somewhat outside the traditional U.N. political structure. Yet, since $350 million per annum is such a modest amount in terms of actual developmental needs, even under the best of circumstances IFAD's contributions can only be marginal. The organization will place new constraints on alternate U.S. courses of action vis-à-vis the fund's beneficiaries. It is difficult to perceive how U.S. food power could be wielded on an adversary basis against the OPEC donors without jeopardizing a useful international mechanism for alleviating the U.S. food aid burden. The International Fund for Agricultural Development is a first step.

The Fertilizer Problem

An assessment of the relative strengths or weaknesses in the food situation of any nation or region should take into consideration the critical role of chemical fertilizers. The production of these fertilizers is closely linked to the energy situation. Vast amounts of energy (fossil fuel energy) are required to produce and transport these vital products. In addition, their production and use are subject to new economic realities resulting from the transfer of wealth from petroleum consumers to petroleum producers. Whether the net position of the United States as a food exporter has been strengthened, weakened, or left unaffected by the new role of fertilizers, they must be considered as a facet of the leverage issue.

Roger Revelle has estimated that by the year 2000 world agriculture will consume 160 million tons of chemically fixed nitrogen annually. Its production will require 250 to 300 million tons of fossil fuel. This is four times the 1974 level and represents roughly 4 percent of present fuel consumption. Estimating from 1976 prices, the cost of nitrogen fertilizer would climb to between $32 and $40 billion; the fossil fuel amount of this cost would range from $15 to $20 billion. These high prices and the likely exhaustion of fossil fuels make it doubtful that energy-intensive agriculture, such as that practiced by the United States, can be continued indefinitely at home, much less extended throughout the developing world.[9]

The heavy energy requirements of fertilizer production could be offset, theoretically, by the conversion of humanly inedible crop residues into methane or alcohol, but this would have to be done on a scale not currently contemplated by the U.S. government. In any event, there is an implicit need for enormous capital investments to meet fertilizer production requirements. In the early 1960s, the United States pioneered breakthroughs in the manufacture of synthetic ammonia, using natural gas as feedstock. As a producer of natural gas (of which we have extensive reserves), the United States enjoys a favorable position in this sector of the fertilizer industry. We also have the benefit of enormous phosphate reserves, second only to Morocco's and larger than those of the Soviet Union. Also, the United States, along with the USSR, Japan, and Europe, share the valuable technical capacity of producing about 80 percent of the world's nitrogen fertilizers. With regard to potash, however, Canada, the USSR, and East and West Germany possess about 94 percent of the world's reserves; the United States has only about 6 percent.[10] The U.S. position thus combines several strengths and a few weaknesses—high technical and productive capacity as well as generous reserves of raw materials on the one hand and heavy alternate demands on energy allocation and some materials scarcities on the other. On balance, the

United States will be in a better position than most nations to deal with the complex problems arising from world food needs and the concurrent demand for chemical fertilizers. But production costs will not necessarily remain as favorable as they are now.

Once again, the question of leverage relates to the OPEC nations, some of which, according to James Grant of the Overseas Development Council, ". . . are flaring more gas than is required today for all existing nitrogenous fertilizer production."[11] Several of the OPEC governments have the available capital and the inclination toward investment in fertilizer production facilities that will be critical to the needs of expanding agriculture in the developing world. But the actual decisions to invest or not to invest will determine how patterns of self-sufficiency or dependence on U.S. foodstuffs evolve. Considering the U.S. corporate role in various OPEC petro-chemical industries, one can imagine a scenario involving the three groups involved in IFAD easier than one in which the so-called food and oil weapons are aimed at each other. James Grant has observed that the fertilizer industry has begun a rapid expansion in response to rising prices. He cites two factors which temper the optimistic outlook: the long lead-time required in the design and construction of new plants, which could cause supplies to remain inadequate until the end of the 1970s; and uncertainties in the investment climate in several OPEC countries, which could cause much of the expansion in productive capacity during the decade to occur in the industrialized countries, where petroleum supplies are limited and prices are high. In fact, Grant notes that there is disagreement as to whether production will meet demand by 1980 in view of the $100 billion investment required and the uncertainty concerning future OPEC supply and pricing policies.[12] The large investment required by the U.S. fertilizer industry may additionally constrain the application of food as a political lever.

Beyond OPEC

Proponents of the food-as-weapon theory share with other observers a very legitimate concern about U.S. dependence on foreign petroleum and on overseas suppliers of other vitally needed minerals. The success of OPEC as a cartel has led to speculation about whether other exporters of raw materials will adopt similar tactics with catastrophic economic results for the Western industrialized countries. Since the raw materials exporting nations outside of OPEC in many instances have concurrent population and agricultural development problems, they would appear to be more vulnerable to counter-leverage in the food sector. Figure 1 illustrates the U.S. minerals balance sheet up to the mid-1970s.

In his Club of Rome study, Mesarovic estimates that by the year 2000 the United States will have to import about 80 percent of its ferrous metals and about 70 percent of its nonferrous metals.[13] These figures suggest that the raw materials needs of the United States extend well beyond the petroleum sphere, yet they do not necessarily indicate that the nation will be in a totally disadvantageous position. In its report of 23 December 1974, the Subcommittee on Foreign Economic Policy of the U.S. House of Representatives sounded a fairly optimistic note on the domestic *potential* for producing raw materials. The report noted that statistics on current and future production are based on mineral "reserves" that can be extracted profitably under present economic, technological, and legal conditions. These reserves constitute only a small fraction of the nation's total mineral "resources" recoverable under different economic circumstances. If resources instead of reserves are tallied, the United States has adequate quantities of ten of the basic thirteen minerals. The exceptions are chromium, tin, and tungsten. Even using the more conservative concept of reserves, the nation should be able to meet anticipated needs for fifty out of sixty-five key substances during the next thirty years. The present large-scale importation of foreign minerals results from

Figure 1. U.S. Minerals Balance Sheet, 1976

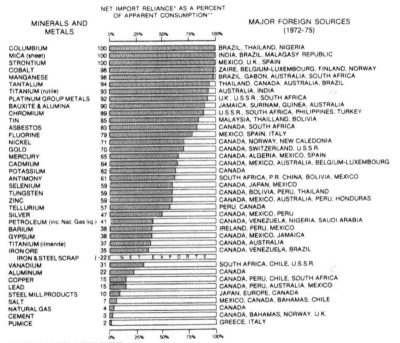

NET IMPORT RELIANCE* AS A PERCENT
OF APPARENT CONSUMPTION**

MINERALS AND
METALS

MAJOR FOREIGN SOURCES
(1972-75)

MINERALS AND METALS	%	MAJOR FOREIGN SOURCES (1972-75)
COLUMBIUM	100	BRAZIL, THAILAND, NIGERIA
MICA (sheet)	100	INDIA, BRAZIL, MALAGASY REPUBLIC
STRONTIUM	100	MEXICO, U.K., SPAIN
COBALT	98	ZAIRE, BELGIUM-LUXEMBOURG, FINLAND, NORWAY
MANGANESE	98	BRAZIL, GABON, AUSTRALIA, SOUTH AFRICA
TANTALUM	94	THAILAND, CANADA, AUSTRALIA, BRAZIL
TITANIUM (rutile)	93	AUSTRALIA, INDIA
PLATINUM GROUP METALS	92	U.K., U.S.S.R., SOUTH AFRICA
BAUXITE & ALUMINA	90	JAMAICA, SURINAM, GUINEA, AUSTRALIA
CHROMIUM	89	U.S.S.R., SOUTH AFRICA, PHILIPPINES, TURKEY
TIN	85	MALAYSIA, THAILLAND, BOLIVIA
ASBESTOS	83	CANADA, SOUTH AFRICA
FLUORINE	79	MEXICO, SPAIN, ITALY
NICKEL	71	CANADA, NORWAY, NEW CALEDONIA
GOLD	70	CANADA, SWITZERLAND, U.S.S.R.
MERCURY	65	CANADA, ALGERIA, MEXICO, SPAIN
CADMIUM	64	CANADA, MEXICO, AUSTRALIA, BELGIUM-LUXEMBOURG
POTASSIUM	62	CANADA
ANTIMONY	61	SOUTH AFRICA, P.R. CHINA, BOLIVIA, MEXICO
SELENIUM	59	CANADA, JAPAN, MEXICO
TUNGSTEN	59	CANADA, BOLIVIA, PERU, THAILAND
ZINC	59	CANADA, MEXICO, AUSTRALIA, PERU, HONDURAS
TELLURIUM	57	PERU, CANADA
SILVER	47	CANADA, MEXICO, PERU
PETROLEUM (inc. Nat. Gas liq.)	41	CANADA, VENEZUELA, NIGERIA, SAUDI ARABIA
BARIUM	38	IRELAND, PERU, MEXICO
GYPSUM	38	CANADA, MEXICO, JAMAICA
TITANIUM (ilmenite)	37	CANADA, AUSTRALIA
IRON ORE	35	CANADA, VENEZUELA, BRAZIL
IRON & STEEL SCRAP	(-22)	NET EXPORTS
VANADIUM	31	SOUTH AFRICA, CHILE, U.S.S.R.
ALUMINUM	22	CANADA
COPPER	15	CANADA, PERU, CHILE, SOUTH AFRICA
LEAD	15	CANADA, PERU, AUSTRALIA, MEXICO
STEEL MILL PRODUCTS	10	JAPAN, EUROPE, CANADA
SALT	7	MEXICO, CANADA, BAHAMAS, CHILE
NATURAL GAS	4	CANADA
CEMENT	3	CANADA, BAHAMAS, NORWAY, U.K.
PUMICE	2	GREECE, ITALY

*NET IMPORT RELIANCE - IMPORTS-EXPORTS
+ ADJUSTMENTS FOR GOV'T AND INDUSTRY
STOCK CHANGES (EXCEPT FOR PETROLEUM
AND NATURAL GAS)

**APPARENT CONSUMPTION - U.S. PRIMARY
+ SECONDARY PRODUCTION + NET IMPORT
RELIANCE PETROLEUM AND NAT GAS
INCLUDES INDUSTRY STOCK CHANGES

Source: Bureau of Mines, U.S. Department of the Interior, <u>Status of the
Mineral Industries, 1977</u> (Washington, D.C.: Government Printing Office,
1977), p. 10.

the fact that the United States has depleted many of its richest and most accessible deposits and finds it uneconomical to exploit secondary reserves at current market prices. Also, the rise of the multi-national corporation has favored transactions of this kind.[14] This economic rationale is not especially comforting from a national security standpoint. In fact, many observers would argue that this dependence on foreign sources constitutes a severe security risk. Food leverage—if this indeed exists vis-à-vis the mineral exporting nations—conceivably could be used to perpetuate short-term economic advantages. At the same time it would hamper the development of domestic minerals reserves.

Further optimistic views of the U.S. raw materials situation may be found in Herman Kahn's *The Next 200 Years* and in the *Models of Doom* critique of the limits-to-growth thesis. Kahn observes that no industrial society can be dependent, in the long run, on any single critical industrial mineral. He believes that few important minerals—perhaps none—will become unduly scarce; however, cartels occasionally might be able to extract extraordinarily high prices and thus necessitate conservation, substitution, and redesign.[15] In *Models of Doom*, the authors point out that natural resources are not so quantitatively limited as man's economic and technological ability to exploit them.[16] The "known reserves" table prepared by the Council on International Economic Policy, reprinted in Kahn's volume, suggests that the spectre of mineral exhaustibility could prompt unnecessary international economic maneuvering as a result of new discoveries.

The problem of the availability of raw materials, like that of the population explosion, is that probabilities of long-term solutions do not necessarily lessen policymakers' burdens during the short-term. The discovery of new minerals, technological advances leading to substitutions, and the continual reordering of commodity values will enable the United States, no doubt, to fulfill its needs over the long-run. But the question remains whether raw materials exporting countries will attempt to form cartels, like OPEC, capable of exerting economic pressure. If such becomes the case, the issue of U.S. food as a viable lever in the quest for imported raw materials becomes even more relevant.

Although the formation of cartels for mineral exportation would indeed complicate the already complex economic problems resulting from oil price increases, economists distinguish between oil and other non-fuel minerals as objects of international competition. Writing in *Foreign Affairs,* Varon and Takeuchi have noted that although demand for most non-fuel minerals is price-inelastic in the short-run, it is far more elastic than oil over the long term. Historical experiences with tin, aluminum, and copper, for example, suggest that the eventual drop in demand for non-fuel minerals would more than offset any price increase, such that the total return to the producer will eventually be less than that before the price change. The reasons for this are threefold: consumer stockpiling, recycling possibilities, and the use of substitutes. Also, the distribution of world reserves cuts across political and economic lines. The developing countries are estimated to have roughly 40 to 45 percent of the world's major non-fuel mineral reserves, 35 percent is in the hands of the Western industrialized nations; and 25 to 30 percent is held by the "centrally planned" (i.e., Communist) countries.[17]

The previously referenced House Subcommittee report on commodity scarcities agrees with the supposition that other producers of raw materials are unlikely to emulate the oil-exporting nations in forming an effective cartel. The mineral industry lacks those characteristics required for the establishment of workable cartels: their sources of supply are more diversified; there is no coalescing political catalyst; substitution opportunities are more available because metals are more interchangeable for many uses; stockpiles serve as a buffer; reserves are on hand in the form of recyclable scrap and lower-grade sources of supply; the lead time to develop a new or alternate supply is less; producers do not enjoy sizeable foreign exchange reserves but rely heavily on exports to pay for imports and to provide employment; some demands are postponable; and finally, the mineral market lacks a special feature of the international oil market, where a few large corporations act as tax collectors and transmission belts for the exporting nations and thus mask the direct confrontation between buyer and seller. The House report cautions, however, that the OPEC nations do not fit the ideal cartel model and yet have been extraordinarily effective in imposing their will on an oil-consuming world. One must realize, too, that the actions of other commodity-exporting nations to control supplies and prices do not have to follow exactly the OPEC pattern. For example, Morocco has utilized its semimonopoly on phosphate to quadruple prices in the past three years, and Jamaica has achieved a sixfold increase in its bauxite prices. The report concludes that the two commodities for which cartel action would seem most feasible are copper and bauxite.[18] It fails to discuss U.S. food in terms of economic counteraction, instead recommending pursuit of U.S. policies that would help assure adequate world food supplies. The apparent reluctance of Congress to consider the possible leverage afforded by U.S. food surpluses and statements made by both presidential candidates during the 1976 campaign underscore the fact that food strategy is a politically sensitive issue. Future policy options relating to employment of food leverage could well be constrained by public and congressional disapproval. These attitudes could change markedly if more serious raw materials scarcities or other economic problems arise.

Other North-South Issues

Occasionally, someone expresses the concern that U.S. efforts to export its agricultural know-how constitute a technology transfer that could reduce the competitive advantage the nation now enjoys. According to this logic, this transfer reduces whatever political and economic leverage the United

States now has. The proposition is not unreasonable. But available evidence suggests that U.S. efforts to expand the green revolution and to share other aspects of our agricultural technology do not appreciably alter worldwide patterns of surplus and deficit. On the contrary, the green revolution has been described by its leading practitioner, Norman Borlaug, as merely a holding action against serious deficits that will occur as the result of population pressures in the future. Assuming that the United States pursues well-designed assistance policies, an increase in agricultural productivity in the underdeveloped nations will merely lessen the overdependence of these countries on U.S. food aid and concessional sales. By contrast, many observers have felt that food aid in the form of Public Law 480 detracted from some of the recipients' efforts to increase domestic productivity and, in some instances, disrupted international markets.

The more developed a country becomes, the greater its potential demand for agricultural products of the type exported by the United States. Many Americans are unaware that the United States is a large-scale importer of foodstuffs. U.S. exports, mainly of grains and other unprocessed food, amounted to $24 billion for the year ending 30 September 1977; imports— for example, tropical products and processed foods—during the same period came to slightly over $13 billion.[19] Just as Americans are willing and able to pay for their quality diet, other nationalities may be expected to trade more actively in the international food markets as they achieve the means to do so. Meat is usually the item showing the most noticeable rise in demand. This could be advantageous for both the U.S. grain and livestock industries.

Land as a Resource

Another determinant of the degree of advantage and hence leverage that U.S. agriculture will enjoy in the international marketplace during the coming decades is the availability of cultivable cropland. Some observers, notably Georg Borgstrom of Michigan State University, have warned that encroaching highways, suburban developments, and other land-conservation projects are taking an alarming quantity of good cropland out of production each year. They feel that this will eventually reduce our ability to feed overseas populations either through trade or aid. On the other hand, several researchers with the USDA's Economic Research Service have produced a brief but well-documented study that concludes that the U.S. land resource base appears ample for domestic needs, at least until the year 2000, and that U.S. farmers will have adequate land and technology to make sizeable commitments to the export market.[20] The measure of performance will be in terms of productivity per crop-acre and the number of currently marginal acres that can be brought into production. Although land apportioned for urban use has doubled since 1950, it still comprises only about 1.5 percent of the land area in the United States. Land withdrawn for transportation purposes subtracts an additional 1 percent. The current base for crop production is about 385 million acres. There has been a downward trend of about one million acres per year during recent decades. This decline has resulted primarily from the retirement of small farms that have proved uneconomical to cultivate with modern machinery. This has been offset by increased productivity as well as declining land needs in some sectors—for example, for cotton. Between 1961 and 1972, land set aside or diverted from crop production under federal production limitation programs ranged from 37 to 65 million acres. With the relaxation and eventual termination of the federal set-aside program in 1973-74, 37 million acres went back into production within a short time. If the weather had been better, 8 million more acres would have been put to use for crops.

Since 1950 U.S. agricultural productivity has risen by 50 percent. This achievement has been attributed to increased mechanization, increased use of improved fertilizers, and various technological advances. The trend can continue, provided agriculture receives priority energy allotments. Alternate sources of nitrogen will be available, and phosphates and potash supplies should be adequate. New hybrid varieties of wheat, barley, and soybeans will soon be available commercially, and nonchemical insect control techniques are being refined.

Technological advances in livestock production have been occurring more slowly. But wider cross-breeding and artificial insemination can be expected to increase beef production by 20 percent within the near future. This increase would allow more acres to be devoted to cereal crop production for human consumption. The increased utilization of forages and crop residues could replace feed grain concentrates as the need arises. A continuing trend toward using synthetics could reduce the land requirements for natural fibers, but this, of course, would depend on the price and availability of petroleum.

Abandonment of economically obsolete cropland is being offset to some extent by the reclamation of new and, in some cases, more productive cropland—at a rate of about 1.3 million acres per year. This results from expanded irrigation, drainage, and land clearing, the reduction of summer fallow, and dryland farming techniques. According to a decade-old USDA land survey cited in the Cotner study, the 365 million acres then used as cropland could be supplemented by cultivating as much as 152 million of the remaining 265 million acres of forest and grassland. The trade-offs would affect the forest and livestock economies. The Cotner study concludes that the U.S.'s land resources have a "sizeable potential" for meeting future domestic and foreign food and fiber requirements, even if present yield assumptions prove overly optimistic.

Cropland availability will continue to be a critical issue because of global demands. Accordingly, new attention will have to be given to various land use issues, to compiling land productivity inventories, to reappraising food demand in the context of higher production costs, and to assessing the feasibility of new technology.[21] In sum, there are definite limits to expanding the U.S.'s croplands, but they should not affect the food leverage question.

The Special Case of the USSR

Any consideration of the potential value of U.S. food as an economic or political bargaining tool would be incomplete without reference to the unusual bilateral relationship that has resulted from the U.S.-Soviet wheat deal. Compared with the less advantageous U.S.-OPEC relationship, this agreement promises to offer a sphere for innovative policies that could give real meaning to the concept of food leverage. Let us look at evolving U.S.-Soviet agricultural relations in terms of past and future opportunities.

The massive Soviet imports in the 1970s were indeed a destabilizing element in the world food situation. Occurring in tandem with the world energy crisis, the purchases and the ex post facto wheat agreement between the United States and the USSR provided important lessons for both countries. Conflicts of interest between U.S. farming and consumer constituencies were brought to light, as well as the vulnerability of the developing world in the interrelated spheres of food, fertilizer, and energy.[22]

Reference was made in Chapter 1 to the fact that the Soviet purchases were made to salvage livestock production plans. Also mentioned was the importance the Soviet regime apparently attaches to offering its people an improved, higher meat-content diet as a sign of a higher domestic living standard. Two factors indicate that the Soviet regime will continue to have a vital interest in U.S. food exports: a standing commitment to improve the Soviet national diet and continuing problems with domestic agricultural production. When one considers the multiyear sales commitment in the present agreement and the prospect of more attractive agreements in terms of selling price, it is safe to predict that the interest of the U.S. farming and commercial constituencies will also remain high. It is difficult to project the food leverage issue into the area of détente, since this political concept remains ill-defined and subject to varying interpretations on both sides. At a minimum, increased bilateral trade in genuinely needed commodities should provide incentives to resolve conflicts in other areas.

The second essential bilateral trade issue is the degree of U.S. need for Soviet exports. Alloy metals such as chromite and bauxite come immediately to mind. The potential for trade-offs between food and oil is much less certain. Future petroleum discoveries in the large land mass of the Soviet Union may well exceed proven reserves by a factor of three or four, according to a published estimate in the Club of Rome study. If this proves correct and if sufficient development is undertaken, the USSR could become a major petroleum exporter in the coming decades.[23] However, the release of an unclassified Central Intelligence Agency (CIA) report in mid-1977 has lent support to the view that the Soviets will be unable to meet export commitments and domestic requirements in future decades unless reserves are brought "onstream" with enormous investments in exploration and development. If this should be the case, petroleum, like food, could be a source for Soviet concern rather than a tradeable asset.

Historically, both the United States and the Soviet Union have demonstrated an inclination to import basic commodities— oil in the former case and food in the latter—for reasons of economic expediency and in order to avoid depleting domestic resources. U.S. oil purchases from OPEC in the post-1972 era most certainly represent a growing dependence on foreign sources. And Soviet food needs could become even more acute during the coming years due to climatic variation and chronic difficulties with agricultural productivity in that country. The trading of U.S. food, computers, and various other sophisticated products for Soviet raw materials may bring enough benefit to both nations to provide a strong rationale for political détente. Thus it is increasingly important that U.S. policymakers come to an accurate assessment of the value of U.S. food in this context; it is equally important that they conclude grain sale agreements in which the food leverage potential is fully realized. One presently unresolved issue broached by the world press in early 1977 relates to a possible increase in world demand for foreign

oil. Competition with the West for oil could alter the food equa-
tion. In the absence of hard information concerning any new
Soviet oil strategy, it is all the more difficult to devise a coherent
U.S. food strategy.

China's food situation is more difficult to analyze than that
of the Soviet Union. Its announcement in March 1977 that as
much as $550 million in scarce foreign-exchange reserves would
be spent on imported grain suggests that it continues to waver
between near self-sufficiency and substantial deficits. Since the
early 1960s, Chinese wheat imports have averaged about 5 to
6 million tons per annum. The 1977 decision may have resulted
from rail transport problems as well as production difficulties.
The implications for future U.S. food leverage are obvious, but
at this stage ill-defined.[24]

Leverage versus Weaponry

Whether U.S. food should be used as a weapon or a lever is
not a question that can be answered with a categoric yes or no.
Food represents many things to many constituencies—farmers,
consumers, organized labor, the foreign trade community. In
the foreign affairs sector some perceive it as a political tool and
others as a humanitarian device. Food simultaneously locks the
United States into binding economic commitments with friends
and allies, incurs humanitarian responsibilities toward the underdeveloped world, provides potential leverage with political
adversaries, and offers theoretical advantages in trading with the
owners of vital raw materials. The United States will probably
continue to enjoy its exceptionally advantageous position as the
world's major food-producing nation, but the attractiveness of
U.S. food as an apparent solution to short-term problems could
divert policymakers from giving due consideration to food in
the context of middle- and long-term challenges. The more
closely one examines the complexities of food as a component
of U.S. foreign relations, the more obvious it is that the judicious

political use of food leverage is a concept that is not incompatible with humanitarianism or with the maximization of national economic interests. Food weaponry is a concept that overlooks real-world complexities and—irrespective of moral or humanitarian considerations—should be avoided because it is infeasible.

5
Policy Priorities and Political Realities

If any message emerges from the preceding chapters, it is that the United States' uniquely productive agricultural system exerts a disproportionate amount of influence on the world food balance. Agriculture seems to be regaining pivotal importance on the domestic economic plane as well as acquiring new importance in both the economic and political areas of U.S. external relations. Clearly, the precipitous energy-related events of the early 1970s have created new economic realities on a world scale. And the United States is growing less immune to them. Food has an enormous role to play in this picture, not as a single issue but as a complex set of interrelated problems.

The Policymaking Environment

Discussion in previous chapters has focused specifically on climate, population, and the economic and political significance of a highly productive national agricultural base in the sphere of foreign aid and trade. These are by no means the only areas in which food policy challenges will arise. Problems confronting the U.S. farmer, the consumer, the food processor, and the national defense establishment are also affected by national food policies. Each area has domestic and international ramifications. Timely and judicious decisions in one area can hardly be relied upon to provide success in another.

U.S. agricultural—and by extension, food—policies traditionally have been pragmatic, limited in scope, and oriented
toward very distinct constituencies. The most easily identifiable
among these constituencies are the farmer, the consumer, the
commercial middleman in the food-processing chain, and various
federal bureaucracies concerned with overseas aid and trade. It
would be unreasonable to expect much, if any, mutuality of
interest among these disparate groups, aside from a vague desire
to see U.S. agriculture contribute ever more efficiently to the
national welfare. What defines a significant contribution and
what constitutes the national welfare are perceived very dif-
ferently, of course, according to the particular vantage point of
each constituency.

The federal bureaucracy in the post-industrial United States
is beset by a host of new social, political, and technical problems associated with the nation's transition into what has been
termed (perhaps optimistically) a *service economy* stage. To
argue that food policy should be assigned a higher priority than
other comparably urgent issues would be unreasonable. One
could argue, however, that food is less well perceived as a long-
term policy problem than some others both inside and outside
government because of the many complex component issues
and the different importance ascribed to these issues by the
various interest groups. Accordingly, national decisions on both
the domestic and international agricultural fronts will probably
continue to be made on an incremental basis and in the absence
of very clear views about the dimensions of the world food
problem and the relationship of U.S. agriculture to this problem.
Although in any nation, political and economic realities of the
moment generally favor short-term, ad hoc policies, this fact of
life does not have to preclude the development of better, long-
term policies. Unless the attention of the public and government
is focused on rationalizing the conflicting interests of geographic, economic, and political sectors in the country, U.S.
agricultural policy will continue to evolve along a traditional
path under decidedly non-traditional global circumstances.

Recognizing that food policies cannot be designed to please all of these groups, at least improved communication and understanding might help in making optimal choices among alternatives. It is apparent, too, that attention will have to be paid to such unfamiliar areas as long-range climate prediction, satellite crop surveys, and population planning. Commendable efforts are being made to understand the extraordinary technical problems involved in feeding the nation and some of the rest of the world; however, it seems that significantly less attention is being directed at the larger food problem—the composite of these myriad mini-problems.

Problems versus Crises

Before assessing some of the key agricultural and food policy issues perceived by the major interest groups in the United States, it is useful to reconsider briefly the question of whether there is an incipient world food crisis. Or is there merely the prospect of continuing fluctuation in the global food balance, causing extreme hardship for a few of the poorest countries and only momentary or periodic concern for more affluent peoples? The possibility exists that some observers have been unduly pessimistic and have overstated their case. In any event, very diverse views exist concerning what will result from the interaction of global weather patterns, population trends, and evolving agricultural technology.

According to one knowledgeable observer of the international situation, there are four essential food facts:

1. There is not one but many food problems, and a surprisingly large number of them are the result of human and governmental decisions rather than immutable forces.
2. Food production can be and has been increased or decreased quite rapidly in "normal" conditions.
3. Food supply and price stability depend largely on stocks

of food large enough to overcome shortfalls in production.

4. When food prices rise sharply, the poor are adversely affected.[1]

The latter two points are sufficiently well understood to constitute truisms, but the first two tend to be overlooked. Interpreted positively, they suggest that mankind has the capacity to deal with food problems that will be at least partially of its own making. The highly regarded Jean Mayer points to another set of factors which he regards as adversely affecting the world food supply and permanently altering the hopeful outlook of the 1960s and early 1970s: (1) population growth; (2) inequality in production and consumption; (3) the diminishing effectiveness of the world fishing industry; and (4) the inadequacy of the green revolution.[2]

Sidestepping the difficult problem of accurate forecasting, one may reasonably assume that adverse trends in the world food balance will be felt most acutely in the underdeveloped countries, where the population pressures are greatest and agricultural productivity is lowest. Accordingly, a scenario involving localized crises rather than a single global one would appear more realistic. According to the 1976 report of the U.N. Rome Conference by the thirty-six-member World Food Council (WFC), the annual grain deficit facing the developing nations will reach 85 million tons by 1985 and will have the most severe impact on the forty-three "food priority" countries, which will be unable to pay for required food imports because of insufficient domestic farm output. This projected deficit is about double that of 1974, when the world came close to running out of food. In the Council's view, food imports by the developing countries have risen alarmingly, their nutritional problems have grown, and yet their production performance has not markedly improved. The staff criticized vested political interests, agricultural lobbies, and outmoded social policies as obstacles to boosting food output. Furthermore, it described the results of the 1974 World Food Conference as "meager," noting that food and fertilizer aid has lagged behind established goals. The staff concluded that the developing nations will have to double the growth of their food production and will require $7 to $8 billion annually in outside loans and aid to prevent "disastrous" shortages in the 1980s.[3]

Another observer of the world food situation takes a less pessimistic but more questioning view. Lyle Schertz suggests that:

> if one takes a measured view of all factors—and if the worst does not happen in weather and energy supplies—it is still possible to visualize an uneasy overall balance between supply and demand for food over the next decade. Yet both will fluctuate substantially from year to year and from area to area. Supply levels are bound to be affected both by weather variation and by the uncertain availability and price of fuel and fertilizer. The level of demand, especially for imports, is highly uncertain. The effect of the fuel crisis on the capacity of nations to pay, and especially the unpredictable decisions of countries such as Russia and China, are involved. Thus . . . there will be wide fluctuations in prices in coming years.[4]

The two interpretations are not conflicting, rather they suggest two problem-areas that will necessitate policy decisions on the part of the United States. The situation envisioned by the WFC report touches on aid issues of a relatively familiar nature which can be handled with at least partial success by familiar bureaucratic mechanisms. But the possibility of widely fluctuating supply and demand presents a more insidious problem because, in this event, the United States would not be cast in the role of a have nation dealing with have-nots. It would have to contend with the impact of fluctuating world agricultural prices on its domestic production base. Herein exists one of the greatest obstacles to the development of a comprehensive, long-range food policy. It is extremely difficult for any administration to focus on hypothetical problems when contemporary problems bear so strongly on the interests of important political and economic constituencies. It is fully understandable that farmers and consumers will be interested (for obverse reasons) in the immediate price situation. Although the several competing foreign relations constituencies presumably are more future-oriented, they too function within a time-frame considerably shorter than that ascribed by knowledgable observers to the evolving world food problem. For example, the foreign aid community must operate within the limits of annual budgets and programs that have no more than a few years' scope. The foreign trade community—spanning both the private and governmental sectors—operates in the very contemporary world of the international marketplace. Solutions to the myriad problems of today have real payoffs, whereas the rewards that might accrue from anticipating the domestic and world agricultural situation a decade or more hence are nebulous at best. These economic and political realities are not good or bad. They simply suggest that it would be unrealistic to call for a simplistic, future-oriented national policy. By the same token, ad hoc solutions to immediate problems are usually more successful if based on public and governmental knowledge of the long-term consequences. Failure in this regard carries high implicit costs. As noted by James Grant of the Overseas Development Council:

> By early 1974, the United States—the food and fertilizer center of the world—was not implementing even the semblance of a global food policy beyond that of maximum profits for agricultural exports, which increased from approximately $9 billion in 1972 to $22 billion in 1974. . . . Only with the World Food Conference did a reversal of these trends begin, and many of the basic policies for the future remain to be decided.[5]

The American Farmer

Since U.S. agricultural policies appear to be more closely linked to near-term market realities than to more abstract and difficult-to-predict problems of international scope, it is essential to begin by examining the present situation of the American farmer. This individual, or to be more exact the American farm family, is usually the forgotten element in the productivity equation that gives the United States such a unique position in world agriculture. Economic realities for the American farmer have immediate political repercussions which can and do outweigh almost all other policy considerations.

Considering his extraordinary efficiency as the prime producer in the food chain, it is ironic that the American farmer currently faces so many short- and middle-term problems. Foremost among these is the increasing capital investment required for a profitable operation. Machinery and land costs have risen enormously, and these are two key productivity elements. While attempts have been made to solve the capital problem through the corporate agri-business approach, various studies have shown that the well-equipped commercial family farm continues to be more efficient than larger, industrial-type food production units. Yet, because of the nature of the prevailing marketing and pricing systems, the traditional farmer does not benefit fully from his productivity gains and is often penalized by excessive production.[6] Ironically, rising land prices do not reflect productivity increases as much as speculative interest in resale. This makes many farmers appear rich on paper and simultaneously drives up leasing prices for additional acreage that must be cultivated to amortize the larger investments in machinery and other fixed assets. In addition, the farmer must borrow at high rates for yearly operating expenses.

The once politically powerful farming constituencies are losing influence in state legislatures and the national Congress (while the urban areas are gaining influence) and, thus, legislative relief in these areas is more difficult to obtain. Once agriculture bills were so assured of passage that they served as vehicles for nonagricultural "riders," yet today the reverse situation prevails. Within the Congress, changes in the seniority and committee systems are working to the detriment of the farm lobby. And at the grass roots level the farm population has a diminishing political voice due to its dwindling size and the shifting of national concerns. Memories of the windfalls created by the crop subsidy and soil-bank programs of recent decades and the domestic price impact of recent foreign grain sales have dampened public sympathy for the farmers' current difficulties. There is, in fact, some truth to the adage that the American farmer is opposed to government interference when food prices are high and is quick to call for government protection when times turn bad. It must be remembered, however, that this phenomenon is not unknown elsewhere in the business and labor sectors. Significantly, it is the vacillation between protectionist and free-market sentiments on the part of the farmer and the federal government both that impedes the development of longer-range policies that would allow U.S. agriculture to serve both domestic and foreign needs. The grain reserves issue provides an excellent example.

The Perennial Problems of Subsidies and Reserves

Since reserves pose a critical domestic farm policy issue and also figure prominently in efforts to solve the world food problem, they merit discussion in both contexts in this concluding chapter. On the domestic plane, it is a fair assumption that all but the most ruggedly individualistic farmers favor some governmental parity scheme to keep their purchasing power in line with that of other economic sectors. This system of governmental intervention for price-support purposes has been time-honored. It dates from the Federal Farm Board of 1929 and the Agricultural Adjustment Acts of 1933 and 1938. According to Hailstones and Mastrianna, the introduction of the parity concept to the farming economy made explicit the social dimensions that had previously been implicit: "The Federal Government committed itself to the philosophy that the farm population should not be subject to an economic situation which placed it in a significantly inequitable position regarding income and purchasing power relative to the rest of the population."[7] The soil-bank program of the post-World War II era was merely a variation on this effort to support income while dealing with the nagging problem of production surpluses. By contrast, the post-1972 era has been characterized by periodic shortages and high prices and resultant "good times" throughout most of the U.S. farm sector. Between 1970 and 1973, government commodity holdings declined 75 percent in value, Commodity Credit Corporation loans dropped by more than 50 percent, and government holdings of specific crops dropped to the lowest levels in decades. However, Hailstones and Mastrianna foresee that agricultural prices will be more susceptible to fluctuations due to a decrease in acreage controls and a depletion of surplus stockpiles. Thus, "a substantial increase in demand, from American or foreign consumers, or a shortfall in production will intensify upward pressures." Conversely, "bumper crops here or abroad can exert downward pressure on food prices. Government food stockpiles, which formerly helped to stabilize the market price . . . are gone."[8]

Parity mechanisms continued to exist on paper during the middle 1970s, but government price-supports were considerably below those in the commercial marketplace. Yet with the recent decline in prices as the result of an upswing in global production, the parity system may find new utility. How it will work under the specific conditions of the late 1970s and what related grain storage schemes evolve will, of course, influence the direction of other food policy developments. One idea espoused by farmers is to have a basic 100-percent parity system in conjunction with a farmer-controlled, on-the-farm strategic food reserve system. Commodities could enter the reserve through a nonrecourse loan program and could be released or called by the government only when the market price reached 100 percent of parity. Farmers would have the option of reselling commodities if prices failed to reach the ceiling level. The proponents of this also favor expanded international trade agreements to stabilize prices and support the concept of an emergency food reserve for international use.[9] Such on-the-farm storage propositions would appear to have the double virtue of stabilizing domestic prices and supporting U.S. actions in both trade and aid abroad.

The Farm Strike of 1977-78

The foregoing problems of farm income, government subsidies, and grain reserves came into public view abruptly with the founding of the American Agriculture Movement in September 1977. This sudden welling of farmer frustration as a result of low market prices, escalating land costs, and equipment debts led, in rapid succession, to the national farm strike, a remedial Administration farm plan, and an abortive farm bill.

The major significance of these events was that the normally individualistic and politically weakened farmers organized so quickly. The strike was significant, not for its limited successes and offsetting failures but for the fact that it occurred at all under these circumstances. The conventional wisdom had been that the burden of individual debt, the consequent need to plant and sell at any price, and a general lack of cohesiveness among farm organizations would make a strike unthinkable.

In retrospect, this crisis may be viewed as something of a blessing in disguise. The public is already more aware of "the farm problem," and the economic and political advantages of increasing U.S. grain reserves have become a little more apparent. The questions remaining are: will market improvements prompt both farmers and government officials to forget the lessons of the winter of 1978 or will the news-making strike increase both public and private realization that the nation needs both short- and long-term agricultural policies that are mutually coherent? Unfortunately, past experience does not invite much optimism. Yet it is safe to predict that temporary cures will become increasingly ineffective in dealing with the farm macro-problem.

Consumerism as a New Force

The farmer-consumer relationship nominally is that of supplier-customer. But it may well be assuming adversary characteristics. Farmer organizations contend that processing, transportation, and retailing account for the lion's share of price increases, and on this point the consumers may be in agreement. In this regard the Food and Agriculture Act of 1977 authorized a new assistant secretary of agriculture position with jurisdiction over food, nutrition, and other consumer-related affairs. Although this is hardly the ultimate answer to growing consumer pressures, it represents a historical departure from the USDA's exclusive advocacy of farmers' interests. The expanding role of the consumer's voice in the USDA context and in other federal regulatory agencies should provide an added incentive for the development of a national food policy.

New environmental concerns could affect the agricultural sector as well. Consumer-farmer differences in this area could become more pronounced. An adversary relationship between farmer and consumer interest groups on the issue of food prices would be misplaced, for the two groups have a common complaint against inflationary pressures elsewhere in the food chain. Presumably, both groups would stand to gain from mutually supportive efforts to obtain a long-range national food storage policy if this could be properly utilized to stabilize prices.

Rationalizing Domestic and Foreign Policy Objectives

It would be an oversimplification to assume that new food and agricultural policies could be developed within the orderly confines of domestic production and consumption, on the one hand, and foreign aid and trade, on the other. The interdependence of the two is extensive. There is a historical tendency to confine domestic farm policy deliberations to the congressional farm block and to USDA, and to focus on the concessional aspects of foreign sales at the expense of vigorous retail promotion of exports. It is reasonable to assume that the new economic realities of the late 1970s will favor the participation of many more federal agencies in the policymaking process; also, greater attention will be paid to the purely commercial aspects of food as a means of offsetting balance-of-payments problems stemming from petroleum imports. In this dynamic and predictably unsettled decisionmaking environment, both the executive and legislative branches will be forced to review the actual efficiency of the U.S. agricultural base, primarily in the context of export earnings. This will involve innovative approaches to energy (and by extension, fertilizers), improved weather and climate forecasting, and adaptation to perceived climatic shifts. The growing problem of water resources, soil conservation, domestic population trends, the special problems of the livestock and fishing industries, the growing importance of the processing and distribution cycle in the food chain, the role of agriculture in defense preparedness, and ongoing research and development funding needs—these will all have to be considered.

Reassessing U.S. Productivity

The strength of U.S. agriculture is derived in large part from the traditional emphasis on applied technology. Assuming that this concern with productivity does not wane, the agricultural base should remain healthy. In *Scientific American,* Earl Heady recently commented:

> If just the unused cropland were now converted to crops, if water were utilized efficiently and if all proved new technologies were adopted, by 1985 the Nation could fully meet all domestic demand and still increase its exports of grain by 183 percent over the record average level between 1972 and 1974. Specifically, corn exports could be increased by 228 percent, wheat exports by 57 percent, and soybean exports by 363 percent. Furthermore, if some of the 264 million acres of potential cropland were brought into production, exports by the year 2000 could be even greater.

Grain exports could be increased even without converting any additional land to crops. If American consumers were to substitute soybean protein for only 25 percent of the meat in their diet, less grain would be needed to feed livestock; by 1985 the United States could export 85 percent more feed grains, soybeans and wheat than it did between 1971 and 1974. Alternatively, if American consumers reduced their total meat intake by 25 percent, the United States could export 103 percent more gain. Or if silage were substituted for 25 percent of the grain that is fed to cattle, American grain exports could be increased by 110 percent.[10]

Implicit in this encouraging prediction is the assumption that management efficiency in the agricultural sector will keep pace with changing times. In his book *The Hungry Planet*, Georg Borgstrom notes that increasing U.S. raw materials dependencies force us to reorient and modify our thinking. "The country is no longer limitless. . . . Now the era of economizing, good management, and true housekeeping takes over."[11]

The first of the previously suggested management concerns—energy—has been discussed in the context of the intensiveness of its use by the agricultural sector and in terms of the relatively minimal leverage that food affords in the international competition for petroleum. It is unnecessary to stress again the important linkage between petroleum and fertilizer production and the role of the latter in U.S. production and the green revolution context.

Two additional factors deserve consideration. Because of recent government and public recognition of the historical underpricing of fossil fuels, whole productive processes will have to be reexamined according to new definitions of efficiency. U.S. agriculture should not be exempt from this reexamination. When it comes to energy usage, the transportation, processing, and distribution portions of the food chain are likely targets for conservation efforts.[12] The second issue is the way in which heavy energy usage in the post-farm portions of the food chain could have an adverse impact in a national emergency. One wonders whether unnecessary energy wastage

occurs through over-reliance on truck transport once the food
has left the farm; whether processes that affect the cosmetic
quality of food more than its nutritional value are justifiable
in an era of growing energy costs; and whether distribution
methods are as efficient as they might be, given the urban patterns and vast geographic distances involved. Defense planners
must recognize that under extreme surge or all-out mobilization
conditions, the U.S. food industry would be required to adapt
its complicated production processes to civilian and military
needs. It would have to compete with the energy, raw materials,
and manpower requirements of other key defense industries.
In both industrial and governmental circles there appears to be
a complacency, probably based on the memories of World War II
performance in the food sector, that is unjustified in the tech-
nologically more complex and more energy-constricted 1970s.
A more sobering precedent for crisis management can be found
in the 1973–74 petroleum boycott and the resultant chaos that
permeated the U.S. and global economies. Keeping in mind
this recent example of how an only marginal reduction in
petroleum supplies wreaked havoc with the national transportation system, decisionmakers should pay considerably more
attention to the question of how the food industry could re-
spond to a national emergency (with comparable or even greater
disruptions in other key industrial sectors).

Competing national quests for petroleum could well be the
cause rather than the result of a national emergency. Planning
efforts are needed, then, in two directions: toward improving
coordination with defense industries, especially with regard to
fuel allocations and manpower- and raw materials-quotas; and
toward developing emergency measures whereby various food
processing and refinement steps could be modified or eliminated
in order to allow for stable production and supply during
inevitable disruptions elsewhere in the industrial base. It would
be realistic to plan for a food supply crisis based on even more
negative and economy-wide events than actually occurred in
early 1974. For better or worse, there is every indication that

American food buying habits will favor more rather than less intensive processing and the purchasing of even more convenience products in the years to come. There will be an unavoidable energy price to pay in peacetime and an even costlier one in the event of a national emergency.

Doing Something about the Weather

Like energy, climatology presents problems that are easier to discuss than to solve. In the latter area, increased coordination between the government and the scientific community is essential. The theoretical problems have already been recognized and appear to be receiving adequate research attention. However, improved prediction techniques remain elusive. One encouraging development is the Department of Agriculture's recent commitment to develop a long-range forecasting system to predict natural disasters such as drought, floods, and prolonged cold spells. Long-range in this instance means one year or more. Using a computer model and information from over 100 years of weather records, the method could give 6–5 odds on the probability of drought, for example, within this time-span. Importantly, this effort symbolizes a new recognition within the executive branch of the need for better weather information.[13] The major challenge for prediction, however, remains in the truly long-range period, during which cooling, warming, or increased variability trends could trigger the previously discussed global shortfalls. These, in turn, would require unpalatable trade-offs in the allocation of U.S. agricultural exports for trade and aid purposes.

Ironically, the goal of improved climate prediction, which is universally endorsed, carries with it potential diplomatic problems. By virtue of its technological achievements, the United States will play a leading role in climatological research and forecasting. The question inevitably arises whether this information should be handled as a rarefied form of commercial or even strategic intelligence or whether it should be fully shared with all countries as part of an international effort to alleviate world hunger. The latter would appear to be the obvious course to follow, but there are problems with it in both the trading sector and the political context of East-West and North-South relations. It is conceivable that other nations might resent and consequently attempt to discredit pessimistic predictions concerning their own climatic future. The question of how far cooperation should extend in cases of adversary political relationships is also problematic. The success of any long-range forecasting would depend to a large degree on cooperative inputs of statistical data. A double standard (some countries might disseminate less data than they receive) obviously would not work. Accordingly, the Bellagio conference's appeal for a world climatic data bank is timely and appropriate.[14] For the United States, the proposed National Climate Program Act represents a quantum leap forward.

The western states' drought of 1977 in the United States has served as a reminder that water is a precious raw material. Georg Borgstrom describes this as a kind of "unsung dimension" to the food problem and warns of the dangers in the profligate use of the earth's groundwater. (This is defined as the subterranean water available within the top mile of the earth's crust; it constitutes one-tenth of all available fresh water resources.) Noting that there is an increased tendency to utilize deep wells for agricultural purposes, Borgstrom does not worry so much about its use as a temporary supplement in times of critical shortage as he does about its regular use as the basis for food production and preparation in an era when humans are counted in billions. Europe presently draws three times more water than is returned to the earth, and North America draws twice as much water as it gets. Borgstrom cites water as an "example in our management of basic resources of how we confuse capital outtake and productivity. . . . Therefore, the water requirements of our daily food constitute not only a key issue but a major neglected sector of our existence which we have taken for granted despite all indications of shortages and excesses."[15]

During the Bellagio conference on climatic change it was suggested that "research on dry farming . . . could be accelerated, and existing dry-farming technology could be applied much more widely to increase crop yields and reduce the chances of crop failures in many semiarid areas that are now under cultivation."[16] This advice is immediately applicable to the developing world, and it may well become appropriate to the United States if predictions of increased drought or climatic variability are borne out.

Conservation Issues

Soil and water conservation are closely interrelated concerns. Despite the lessons of the 1930s and strong governmental support for (and past farmer acceptance of) soil conservation techniques, we cannot take for granted that these techniques will be followed in the future. Pointing to the "surging world grain prices in the mid-1970s, and consequent all-out production efforts by U.S. farmers in 1974 and 1975," one expert has concluded that inadequate conservation practices have raised the spectre of a new dust bowl. This concern is supported by government surveys, which show that of the nearly 4 million hectares of former pastures, woodlands, and idle fields converted to crops nationwide in late 1973 and 1974, over one-half were inadequately handled.[17] The mid-1970s urge to till every available inch of Midwestern land (which is the most productive), has resulted in the cutting of many trees that were planted as windbreaks during the conservation projects of the late 1930s. The Dust Bowl episode occurred within memory of all Americans now of middle age or older, and yet it appears that the intervening thirty years have erased appreciation of the causal factors.

Fish traditionally have played a minor role in the typical American diet. Consequently, there has been a tendency to overlook fishery resources and the condition of the fishing industry as elements of the national agricultural base. This tendency may be reversed as a result of the imposition of a protective 200-mile limit by the United States. Judging by the seizure of Soviet vessels that have violated catch limits and the generally increased surveillance and contingency planning by the U.S. Coast Guard, these restrictions will be realistically enforced. Increased governmental attention to law-of-the-sea matters and this belated appreciation of the dangers of overfishing in continental waters by foreign fleets obviously are steps in the right direction. But the U.S. fishing industry remains grossly underdeveloped with regard to the economic and nutritional value of this food resource. The new activist policies represent an encouraging development on the side of conservation. Yet at some future point, presumably within a decade, conservation will have to be balanced by a more creative exploitation. National dietary standards cannot change overnight, but interest in fish could change—or be changed—as a result of cost reductions through more efficient fishing methods, an adverse shift in beef prices, and possibly a greater awareness on the part of the public concerning the nutritional advantages of fish. Recent interest in exploitation of seabed mineral deposits within the continental shelf areas which are roughly coincident with the 200-mile limit has focused policymakers' attention on new international legal problems. As a pace-setting nation in technology and a heavy user of minerals, the United States should make major strides in this area. Although the demand for fish is not commensurate with the demand for minerals, the fishing industry should benefit from the technological and the political-legal developments in this area.

From the national point of view, fish may become something more than a marginal protein resource. In the event of worst-case future global famine, advances in marine biology may permit the development of sea-farming techniques, SCP (single cell protein) culture, algae cultivation, and other futuristic solutions to mass-feeding requirements. The enormous expanse of U.S. territorial waters within the temperate climatic zone, especially as fixed by the new claims, could become an asset of value proportionate to that of the fertile, continental land mass.

The Meat versus Grain Debate

The position of the meat industry within the U.S. agricultural base poses a number of vexing policy questions. The cattle raisers' current problems involving high production costs and slim or nonexistent profit margins constitute a touchy economic (and by extension, political) issue. As a result of rising incomes at home and elsewhere in the industrialized world, industry should find more favorable conditions in the future. Prosperity and the demand for more and better meat appear to be intertwined. This relationship can be seen in those developed countries that have advanced rapidly, such as Japan and certain European countries.[18] It is most noticeable in developing countries as per capita incomes first begin to rise. We are reminded that demand for livestock feed could run headlong into grain requirements for human consumption—especially in the less-developed areas where demographic problems are the greatest. As discussed previously, economic differences between the have and the have-not nations quickly assume political overtones. The appeals that some futurists are making to curtail livestock production in the interest of having more grain for human consumption appear to be premature, if not incorrect, by virtue of a misunderstanding of the economics of meat production. As noted by Professor Arnold Paulsen of Iowa State University, a reduction in meat consumption in the United States would not allow the world to eat any better. In the United States, the products of about 600 million acres of grassland and 150 million acres of cropland are consumed in the form of meat and milk. Most of the grasslands are too dry, too steep, or too rocky to be planted with crops economically. We obtain more food from these lands by feeding the vegetation to animals than we would by not using the vegetation at all. Even though the calories in meat amount to only about 5 percent of the calories in the grass, feeding this inedible vegetation to animals increases the total U.S. food supply by the equivalent of 5 million tons of wheat annually.

Viewed over the long-term, livestock serve as a balance wheel in the man-to-food relationship. When the capacity to produce food is limited relative to demand, man keeps fewer meat animals and raises plants which he can eat directly.[19] This economic truism would appear to apply equally well for both long- and short-term global food requirements. In any event, hypothetical, global food problems should not provide an excuse for policy neglect of the distressed U.S. livestock industry. As noted in a recent USDA review, feed-grain policy needs to be shaped so it is able to adjust to uncertainties in domestic and foreign demand. In order to deal with short-run grain surpluses and shortages and allow for a reasonably stable domestic food policy for long-run demands, a good deal of flexibility is required. This would also help avoid the price swings that are so damaging to the industry.[20]

On the technical level, there are additional arguments refuting futurist claims that meat consumption is detrimental in terms of the potential trade-offs in edible grains. Recent British investigations of vegetable protein suggest that genuine animal meat cells contain vital extras in the structural fats of the phospholipid group that are essential for the development of brain and central nervous tissue and that cannot be obtained from plants or synthesized by man.[21] A scientific breakthrough in the area of vegetable protein would have important positive implications for the global food situation, but it is doubtful that the impact on the U.S. livestock industry or the dietary habits of the wealthier nations would be very pronounced. If anything, the likelihood of such a breakthrough occurring within a decade or two refutes those who claim that the current U.S. emphasis on livestock production is wasteful because of the rising demand for edible grains. The consumption of meat in the richer regions and of meat substitutes in the poorer regions would be determined by changing economic circumstances.

New Perspectives

The foreign agricultural policy of the United States traditionally has been a complex issue, and there is every reason to assume that it will become even more so in the final two decades of this century, for energy and raw material constraints, population problems, and other challenges are forcing a reordering of relations among nations. In the post-World War II era, the international relations of the United States were driven by political more than economic forces as a result of seemingly pervasive East-West tensions. The 1972–73 energy crisis appears to have ushered in a new era in which East-West tensions will not disappear but will assume a less central role. Energy and resource problems have been latent in the world scene for many decades, but the precipitous events of 1972–73 apparently were required to bring the issues into focus. Just as it would be an oversimplification to speak of oil as the single problem of today, it would be specious to argue that food will be the overriding problem for an increasingly populous globe. Growing interdependency among nations is matched by a comparable interdependency among political and economic issues. Food, energy, population, climate, and many other subject-areas simply cannot be viewed in isolation. Accordingly, U.S. foreign agricultural policy in the ensuing decades will have to evolve in a less discrete issue environment.

Additional Thoughts on Food Leverage

Like the political and economic components of other international problems, the politics and economics of U.S. food cannot be considered separately. Whether through sale, barter, or gift, success or failure of U.S. food-export programs will be measured in both political and economic terms. Considering the quantities normally available for export and the fluctuating but generally great world demand for foodstuffs, the issue of food leverage will arise regularly during the remainder of the century. In times of increased political tension or other resource constraints, the term *weapon* no doubt will be used instead of *lever*. As suggested in an earlier chapter, the latter concept may be a viable one but the concept of overt weaponry remains an overly simple approach to a very complex issue-area.

As William Schneider suggests, there are already legislative bases for export controls that could be applied to food—for example, the Trading with the Enemy Act of 1917 and the Export Administration Act of 1969. As a proponent of a more aggressive use of food in the international relations arena, he is careful to point out that economic warfare is a device for significantly influencing the behavior of potential adversaries rather than a device for "bringing an opponent to heel" (in the World War II sense of the term). Leverage, he suggests, could be applied by one of two mechanisms—export controls or a national grain reserve manipulated primarily as an instrument of U.S. foreign policy.[22] The former method appears to have been used already in a halting and indirect fashion. For example, the United States denied a request by the Allende government of Chile for wheat sales on credit. The 1972–73 Soviet sales represented a governmental decision in the opposite direction; in this instance it is difficult to judge the relative weight of the economic and political motivations. In the June 1973 soybean boycott, the lever was pushed forward and then back in a rather unsubtle manner.

Several conclusions can be drawn from these and other examples of applications of food leverage. The first and most obvious is that pragmatism and professed national ideals can at times be conflicting. Food is comparable to human rights as a sensitive political issue. Despite the fact that U.S. Public Law 480 programs were motivated by food surpluses and storage problems and other, pragmatic, political considerations, this food aid was regarded by the public as a humanitarian gesture. U.S. food aid programs most certainly should be considered as a significant humanitarian effort and, in fact, probably have received taxpayer (and congressional) support for this reason. It is questionable, however, if this support would have been forthcoming if only the domestic surplus or foreign political factors were stressed. Considering the depth of public feeling about the humanitarian value of food aid and the widespread congressional attitudes reflected in the 1976 effort to pass "Right to Food" legislation, it is even more doubtful that executive branch efforts to utilize food leverage on a new and narrowly opportunistic basis would succeed. On the other hand, passage of new foreign aid legislation in August 1977 has led to an active examination of human rights practices in recipient countries. This could be termed benevolent leverage.

The second conclusion that can be drawn from past efforts to manipulate food for economic and political advantage is that the United States is bureaucratically less well equipped to regulate its food exports than are other countries with national marketing boards or other comparable control mechanisms. The relatively cumbersome nature of U.S. export-control regulations, and the well-entrenched private grain-trading system that operates in both domestic and international markets with considerable independence (if not secrecy), offer numerous impediments to efficient governmental control of trade. Export controls could be revised in the legislature, but it is questionable whether the existing system of private grain trade could be drastically altered.

In his assessment of the limits to food leverage, the USDA's Joseph Willett suggests the following negative factors: alternate foreign export sources; the high storage costs associated with the intentional withholding of exportable surpluses; domestic (U.S.) demand as a constraint on prices and availability of foodstuffs; the sporadic nature of Soviet demand; the fact that Soviet requirements relate to livestock feeding rather than human feeding; the rising U.S. need to export food to recoup payments for petroleum imports; and rising world food production.[23] These factors obviously do not rule out pragmatic U.S. policies regarding food exports, but they do suggest that the leverage strategy is not as promising as its proponents claim.

Bearing in mind the domestic constraints—political, bureaucratic, and attitudinal—that work against the food-as-weapon strategy, policy changes will have to be incremental and not oblivious to human factors.[24] This does not mean that better bargains cannot be struck. Food offers extraordinary negotiating opportunities for the United States and the Soviet Union outside the difficult arena of arms limitation agreements. Indeed, bilateral food trade agreements could expand mutual advantages and propel agreements in other substantive areas. The possibilities of regularized grain trade with China have yet to be fully explored. Although the limitations on food leverage with the OPEC nations have been discussed already, this too is another area where cooperation—especially with regard to third-country assistance efforts and the expansion of global fertilizer production—could provide both economic and political profits. Imaginative approaches are required to shape future U.S. food trade with the developing countries. As Georg Borgstrom has observed:

> The outmoded way of dividing the world into raw material-producing countries versus industrialized nations charged with upgrading these commodities, needs to be thoroughly revised. There is every reason to believe that U.S. agriculture with its food producing potential will emerge within the near future as the nation's most important economic asset. Conversion industries which could mass produce such items as cheap yeast, dried milk, and soybean products, as well as meat, egg, and poultry factories . . . would have their place in such a development. These commodities would be exchanged for cheap cars, metal ores, artificial fibers, aluminum products, and so on. A new, more realistic approach is required, throwing overboard conventional economic patterns which are not only outdated but also serve the world poorly.[25]

Toward More Rational Aid Programs

There is a need to reconsider how U.S. aid policies for developing nations might be revised so as to enhance world agricultural productivity and reduce the demand for U.S. food assistance (which the United States provides at the expense of export sales). In many instances U.S. aid policies have been oriented toward rural development, but the record here is not one of total success or consistency. As Lester Brown has noted, direct food contributions in place of in-country agricultural development efforts have tempted recipient nations to postpone agricultural reforms, have depressed local agricultural prices, and have disrupted third-country export markets.[26] A de-emphasis on direct food aid and increased emphasis on local productivity, however, should not entail undue reliance on the green revolution. Even the revolution's leaders caution that new seed and cultivation technologies can only buy time. Productivity can be outstripped by population. Accordingly, rural development must be linked to more comprehensive, family planning programs. The scope of future aid programs will thus be considerable. In the next twenty-five years, Latin America, Asia, and Africa might require $700 billion for irrigation and agricultural development (i.e., $30 billion per year). This is less than 1 percent of the present gross world product. A large part of the capital investment would have to come from the developed countries.[27]

The Unavoidable Issue of Reserves

Mention has already been made of the importance of creating a new innovative grain reserve and price support system as essential elements in sustaining the domestic strength of the U.S. farm economy. The grain reserves issue looms large in any consideration of U.S. foreign agricultural policy as well—so large that it has received extensive attention in recent publications. The disappearance of the traditionally large government-held grain reserves as an outcome of the 1972-73 massive Soviet grain deal and concurrent heavy export sales to other foreign customers has troubled many responsible observers, who fear the consequences of new famine situations if stocks are depleted. These concerns have been muted only by the occurrence of generally favorable world crop yields during the past two years.

Applied both to the United States and the world at large,

reserves provide the dual advantage of providing market stability and a means of dealing with regional emergency shortages. The one objective is economic and the other is humanitarian, but the two are by no means incompatible. As Fred Sanderson has observed, during the 1960s and '70s, "the world's exportable grain reserves were carried almost exclusively in North America. . . . What is needed is an international agreement by which this responsibility would be shared by both exporting and importing countries." Sanderson notes that Soviet participation would be essential, but that there would be no need for an international bureaucracy to manage a world stabilization reserve, which could be stored in existing facilities.[28]

Another Brookings observer, Phillip Trezise, has noted,

> The most sensitive issues in providing a famine emergency stock probably would relate to its management as a genuine reserve. At one extreme is the possibility that it would be treated as a new variety of food aid to be released to offset the low levels of nutrition prevailing in the poorest countries. At the other might be a reluctance on the part of donor countries to release stocks in their possession, even in the face of an apparently clear and present famine threat; this might be the case, for instance, if cereal prices were rising fast and if domestic political pressures called for action to check their rise.[29]

Addressing the market stabilization side of the problem, Trezise contends that adequate reserves would reduce losses to the contributing countries resulting from sizable market disturbances and that long-run market stability could be the basis for a more nearly optimum use of the world's agricultural resources. The United States and Canada could permit surpluses to develop but could hold and use the resulting stocks to help keep domestic prices stable, giving priority to price stability in their internal markets by limiting (i.e., licensing) allowable exports of grain.[30]

A January, 1977 USDA analysis of grain reserve problems contains several attractive proposals that could provide helpful guidance in the process of creating a new reserves policy. Among these conclusions are: commodity prices are preferable to quantities in establishing the rules for managing reserve stocks; a grain reserve would cost about $300 to $500 million annually if the U.S. Government purchased and sold the grain; the price rules would need to be adjusted over time because of initial miscalculations, inflationary trends, or long-run changes in supply and demand; management rules would need to be "hard and fast" to achieve market stability; and the reserve could be designed to allow farmers to retain ownership (and physical possession) of the grain. Farmers would forfeit their right to sell until the price exceeded a specified level or until the contract expired. Alternatively, farmers could grow grain on designated (set-aside) acreage and store it on farms until needed.[31]

Whatever technical solutions evolve, it is apparent that the dismantling of the U.S. grain reserve system in the early 1970s has created an institutional vacuum that must be filled. U.S. action to establish a new reserve and agricultural price support system should run in tandem with more vigorous multi-national efforts to create a viable world reserve. In the absence of such action, a basically undesirable situation will persist in which the only meaningful global reserves will continue to be confined to North America.

Setting Policy Priorities

Innovations in foreign assistance programs and the re-creation of a viable reserves system typify the middle-term policy challenges of the world food problem, whereas the spectre of widespread famine in an intolerably populous world seems more remote and therefore amenable to further research. Policymaking in the United States traditionally is most effective when the issues are precise and the time-frame immediate. We are a nation of doers rather than planners, and it will be especially difficult in the final two decades of the twentieth century to have to contend with so many global problems of a long-term and abstract nature. Food clearly belongs in this category. Various political constituencies in the United States will continue to have different perceptions and distinctive goals concerning the nature and magnitude of the world food problem. But no constituency can deny that U.S. food is a global resource. There is reason to hope that the value of this resource will be increasingly appreciated as both the executive and legislative branches of the U.S. government are forced to examine new realities in such interrelated issue-areas as energy, raw materials requirements, and balance-of-payments deficits.

Notes

Chapter 1

1. Mihajlo Mesarovic and Eduard Pestel, *Mankind at the Turning Point: The Second Report to the Club of Rome* (New York: E. P. Dutton and Co., 1974), p. 21.

2. Lester R. Brown, *By Bread Alone* (New York: Praeger Publishers, 1974), p. 6.

3. Ibid., p. 44.

4. Sterling Wortman, "Food and Agriculture," *Scientific American*, September 1976, p. 32. The 100 million ton forecast could be lowered to 75 or 80 million tons if several developing countries continue to produce grain in exportable quantities. It must be noted also that the growing import gap can be attributed in part to the increasing affluence among certain developing nations who have ample foreign exchange reserves obtained from petroleum and other lucrative exports.

5. Earl O. Heady, "The Agriculture of the United States," *Scientific American*, September 1976, p. 107.

6. U.S. Department of Commerce, Bureau of Economic Analysis, *Survey of Current Business* 57 (July 1977): S-22.

7. Georg Borgstrom, *The Hungry Planet: The Modern World at the Edge of Famine* (New York: Macmillan Co., 1972), pp. 410-12.

8. Ibid., p. 424.

9. Ibid., pp. 409-10.

10. U.S. Department of Agriculture, Economic Research Service, *The World Food Situation and Prospects to 1985*, Foreign Agricultural Economic Report no. 98 (Washington, D.C.: Government Printing Office, 1974), p. 53.

11. U.S. Congress, House Committee on International Relations, *The Right to Food Resolution, Hearings before the Subcommittee on International Resources, Food, and Energy, on H. Cong. Res. 393*, 94th Cong., 2d sess., 1976, p. 3.

12. See Gordon Bridger and Maurice de Soissons, *Famine in Retreat? The Fight Against Hunger: A Study and A Strategy* (London: J. M. Dent and Sons, 1970), p. 54.

13. Ibid., p. 29.

14. Brown, *By Bread Alone*, pp. 63-64.

15. Lyle A. Schertz, "World Food: Prices and the Poor," *Foreign Affairs* 52 (April 1974): 513, 518.

16. Brown, *By Bread Alone*, p. 4.

17. *Climate Change, Food Production, and Interstate Conflict—A Bellagio Conference, 1975* (New York: The Rockefeller Foundation, 1976), p. 31.

18. Ibid., p. 34.

19. Ibid., pp. 21-22.

20. Everett S. Lee, "Population and Scarcity of Food," *Annals of the American Academy of Political and Social Science* 420 (July 1975): 2.

21. H.S.D. Cole et al., eds., *Models of Doom* (New York: Universe Books, 1973), p. 172.

22. Herman Kahn, William Brown, and Leon Martel, *The Next 200 Years—A Scenario for America and the World* (New York: William Morrow and Co., 1976), pp. 111-12.

23. Brown, *By Bread Alone*, p. 39.

24. According to UNFAO estimates, approximately two-thirds of anticipated increases in food demand will result from population. Rising affluence will account for one-third of these increases.

Chapter 2

1. The words *climate* and *weather* are frequently (but incorrectly) considered synonymous. To be more precise, *weather* describes meteorological phenomena of less than two weeks' duration and *climate* relates to any longer-range phenomena.

2. Henry B. Arthur and Gail L. Cramer, "Brighter Forecast for the World's Food Supply," *Harvard Business Review,* May-June 1976, p. 163.

3. *Climate Change, Food Production, and Interstate Conflict—A Bellagio Conference, 1975* (New York: The Rockefeller Foundation, 1976), pp. 20-21.

4. Deborah Shapley, "Crops and Climatic Change: USDA's Forecasts Criticized," *Science* 193 (September 1976): 1223.

5. If the USSR proves to be a food-deficit nation over the long-run, it will be competing—with superior purchasing power—in world markets against the poorer Third World nations, whose deficit status is already clearly established. See Henry S. Bradsher, "U.S. Experts See Soviet Grain Shortfall," *Washington Star,* 17 November 1976, p. A-4.

6. U.S. Central Intelligence Agency, *Potential Implications of Trends in World Population, Food Production, and Climate,* Report no. OPR-401, August 1974, pp. 27-30.

7. Reid A. Bryson, "The Lessons of Climatic History," *Ecologist,* July 1976, pp. 208-10.

8. U.S. Central Intelligence Agency, *Population, Food Production and Climate,* pp. 30-33.

9. Bryson, "Lessons of Climatic History," p. 205.

10. Lowell Ponte, *The Cooling* (Englewood Cliffs, N.J.: Prentice-Hall, 1976), pp. 29-30.

11. Stephen H. Schneider, *The Genesis Strategy: Climate and Global Survival* (New York: Plenum Press, 1976), pp. 15-16.

12. *Climate Change,* pp. 8-10.

13. John Gribbin, *Forecasts, Famines, and Freezes: Climate and Global Survival* (New York: Plenum Press, 1976), pp. 115-16.

14. See also Lester R. Brown, *By Bread Alone* (New York: Praeger Publishers, 1974), p. 68.

15. U.S. Committee on Climate and Weather Fluctuations and Agricultural Production, *Climate and Food: Climatic Fluctuation and U.S. Agricultural Production* (Washington, D.C.: National Academy of Sciences, 1976), p. 145.

16. Ibid., pp. 145-46.

17. Shapley, "Crops and Climatic Change," pp. 1222-23.

18. The National Academy of Sciences recently published two additional studies that are highly relevant to the problems discussed here. They are: *Understanding Climatic Change, A Program for Action* (1975), and *Climate and Food* (1976). Another key reference work is the Global Atmospheric Research Program's *The Physical Basis for Climate and Climate Modelling*, Publication no. 16.

19. U.S. Congress, House, Remarks by Representative George E. Brown (California) on the National Climate Program Act of 1977, *Congressional Record*, 95th Cong., 1st sess., 4 January 1977, pp. H61-62.

Chapter 3

1. Henry B. Arthur and Gail L. Cramer, "Brighter Forecast for the World's Food Supply," *Harvard Business Review*, May-June 1976, p. 166.

2. Herman Kahn, William Brown, and Leon Martel, *The Next 200 Years—A Scenario for America and the World* (New York: William Morrow and Co., 1976), pp. 27-30.

3. Ibid., pp. 32-34.

4. Ansley J. Coale, "The History of the Human Population," *Scientific American*, September 1974, p. 51.

5. Paul Demeny, "The Populations of the Underdeveloped Countries," *Scientific American*, September 1974, pp. 149-50.

6. Charles F. Westoff, "The Populations of the Developed Countries," *Scientific American*, September 1974, p. 115.

7. See Roger Frisch and Roger Revelle, "Nutrition and

Childbearing Age," *Archives of Diseases of Childhood* 46, no. 249 (October 1971):695-701.

8. Thomas Poleman, "World Food: A Perspective," *Science* 188 (May 1975):514.

9. U.S. Agency for International Development, *The Population Explosion: A Present Danger*, Washington, D.C., 1975.

10. James P. Grant, "Food, Fertilizer, and the New Global Politics of Resource Scarcity," *Annals of the American Academy of Political and Social Science* 420 (July 1975):14.

11. Kahn, *The Next 200 Years*, p. 114.

12. Lester R. Brown, *Population and Affluence: Growing Pressures on World Food Resources* (Washington, D.C.: Overseas Development Council, 1973), p. 9.

13. Donella H. Meadows et al., *The Limits to Growth* (New York: Universe Books, 1972), p. 113.

14. Lester R. Brown, *By Bread Alone* (New York: Praeger Publishers, 1974), p. 180.

15. Ibid., p. 181.

16. Mihajlo Mesarovic and Eduard Pestel, *Mankind at the Turning Point: The Second Report to the Club of Rome* (New York: E. P. Dutton and Co., 1974), pp. 78-80.

17. Ronald Freedman and Bernard Berelson, "The Human Population," *Scientific American*, September 1974, pp. 34-35.

18. Ibid., p. 37.

Chapter 4

1. John S. Steinhart and Carol E. Steinhart, "Energy Use in the U.S. Food System," *Science* 184 (April 1974):313.

2. Ibid., pp. 309-10.

3. David Pimentel et al., "Food Production and the Energy Crisis," in Philip H. Abelson, ed., *Food: Politics, Economics, Nutrition, and Research* (Washington, D.C.: American Association for the Advancement of Science, 1975), p. 125.

4. Ibid., pp. 125-26.

5. Steinhart and Steinhart, "Energy Use," pp. 213-15.

6. Mihajlo Mesarovic and Eduard Pestel, *Mankind at the Turning Point: The Second Report to the Club of Rome* (New York: E. P. Dutton and Co., 1974), p. 22.

7. Dan Morgan, "Our Newest Weapon: Food," *Saturday Review*, 13 November 1976, p. 8.

8. William Steif, "OPEC Welshes on a Promise," *Saturday Review*, 13 November 1976, p. 14.

9. Roger Revelle, "The Resources Available for Agriculture," *Scientific American*, September 1976, p. 168.

10. Lester R. Brown, *By Bread Alone* (New York: Praeger Publishers, 1974), p. 124.

11. James P. Grant, "Food, Fertilizer, and the New Global Politics of Resource Scarcity," in *Annals of the American Academy of Political and Social Science* 420 (July 1975):17.

12. Ibid., p. 18.

13. Mesarovic, *Mankind at the Turning Point*, pp. 26-27.

14. U.S. Congress, House Committee on Foreign Affairs, *Global Commodity Scarcities in an Interdependent World, A Report by the Subcommittee on Foreign Economic Policy*, 93d Cong., 2d sess. (Washington, D.C.: Government Printing Office, 1974), p. 18.

15. Herman Kahn, William Brown, and Leon Martel, *The Next 200 Years—A Scenario for America and the World* (New York: William Morrow and Co., 1976), pp. 86-87.

16. H.S.D. Cole et al., eds., *Models of Doom* (New York: Universe Books, 1973), p. 37.

17. Bension Varon and Kenji Takeuchi, "Developing Countries and Non-Fuel Minerals," *Foreign Affairs* 52 (April 1974):505-08.

18. Committee on Foreign Affairs, *Global Commodity Scarcities*, pp. 6-7.

19. U.S. Department of Commerce, *Survey of Current Business*, October 1977, pp. S-22, S-23.

20. See M. L. Cotner, M. D. Skold, and O. Krause, *Farmland: Will There Be Enough?* U.S. Department of Agriculture Bulletin no. ERS-584, May 1975.

21. Ibid., pp. 3-14.

22. See George E. Brandow, *Impact of Russian Grain Purchases on Retail Food and Farm Prices and Farm Income in the 1975 Crop Year, A Study Prepared for the Use of the Joint Committee of the U.S. Congress* (Washington, D.C.: Government Printing Office, 1975).

23. Cole et al., *Models of Doom*, p. 96.

24. *New York Times*, 10 March 1977, p. 48.

Chapter 5

1. Harry Walters, "Difficult Issues Underlying Food Problems," *Science* 188 (May 1975):528.

2. Jean Mayer, "The Dimensions of Human Hunger," *Scientific American,* September 1976, pp. 45-48.

3. *Washington Post,* 16 June 1976, p. A-2.

4. Lyle A. Schertz, "World Food: Prices and the Poor," *Foreign Affairs* 52 (April 1974):523.

5. James P. Grant, "Food, Fertilizer, and the New Global Politics of Resource Scarcity," *Annals of the American Academy of Political and Social Science* 420 (July 1975):19-20.

6. Minnesota Farmers Union, "1977 Policy Statement Adopted by the 35th Annual Convention" (21-23 November 1976), p. 3.

7. Thomas J. Hailstones and Frank V. Mastrianna, "Food— What Happened to the Surpluses," in *Contemporary Economic Issues and Problems,* 4th ed. (Cincinnati: South-Western Publishing Co., 1976), pp. 169-70.

8. Ibid., p. 175.

9. Minnesota Farmers Union, "1977 Policy Statement," pp. 9-10.

10. Earl O. Heady, "The Agriculture of the U.S.," *Scientific American,* September 1976, pp. 107-27.

11. Georg Borgstrom, *The Hungry Planet: The Modern World at the Edge of Famine* (New York: Macmillan Co., 1972), pp. 422-23.

12. See U.S. Department of Agriculture, Economic Research Service, *Agricultural Outlook*, May 1977, p. 16.

13. *Washington Star*, 7 February 1977, p. 1.

14. *Climate Change, Food Production, and Interstate Conflict—A Bellagio Conference, 1975* (New York: The Rockefeller Foundation, 1976), pp. 18-19.

15. Georg Borgstrom, *Too Many: A Study of the Earth's Biological Limitations* (New York: Macmillan Co., 1969), pp. 143-44, 154-55.

16. *Climate Change*, p. 26.

17. Erik P. Eckholm, *Losing Ground: Environmental Stress and World Food Prospects* (New York: W. W. Norton and Co., 1976), p. 52.

18. Roger Revelle, "Food and Population," *Scientific American*, September 1974, p. 163.

19. Iowa State University, *The Iowa Stater* 3, no. 7 (March 1977).

20. Ronald A. Gustafson, "Livestock-Grain Interdependence: Implications for Policy," in *Agricultural Food Policy Review*, U.S. Department of Agriculture Report no. ERS AFPR-1 (Washington, D.C.: Economic Research Service, 1977), p. 129.

21. John Laffin, *The Hunger to Come* (London: Abelard-Schuman, 1971), p. 221.

22. William Schneider, "Can We Avert Economic Warfare in Raw Materials?" *Agenda*, Paper no. 1, 1974, pp. 11, 17, 31.

23. Joseph W. Willett, "Food as a Factor in U.S.-USSR Relations," *National Security Affairs Forum*, Spring/Summer 1976, pp. 33-37.

24. The International Development and Food Assistance Act of 1975 (PL 94-161) is noteworthy in that it amends the PL 480 Food-for-Peace program to provide better and more effective distribution of food aid abroad and lessens the amount of Title I food that can be allocated for political purposes.

25. Borgstrom, *The Hungry Planet*, p. 422.

26. Lester R. Brown, *By Bread Alone* (New York: Praeger

Publishers, 1974), p. 230.

27. Roger Revelle, "The Resources Available for Agriculture," *Scientific American*, September 1976, p. 178.

28. Fred H. Sanderson, "The Great Food Fumble," *Science* 188 (May 1975):508.

29. Philip H. Trezise, *Rebuilding Grain Reserves—Toward an International System* (Washington, D.C.: The Brookings Institution, 1976), p. 25.

30. Ibid., p. 27. The complex problem of grain reserve sizing has been treated in considerable detail and according to various mathematical and statistical techniques in *Grain Reserve Sizing: A Multiobjective and Probabilistic Analysis* (Austin, Texas: Lyndon B. Johnson School of Public Affairs, University of Texas, 1976).

31. Thomas A. Miller and Alan S. Walter, "An Assessment of Government Programs that Protect Agricultural Producers from Natural Risks," in *Agricultural Food Policy Review*, U.S. Department of Agriculture Report no. ERS AFPR-1 (Washington, D.C.: Economic Research Service, 1977), pp. 91-92.

Bibliography

Books and Reports

Alpert, Paul. *Partnership or Confrontations? Poor Lands and Rich.* New York: The Free Press, 1973.

Bloomfield, Lincoln P., and Irirangi, C. *The U.S., Interdependence and World Order.* New York: Foreign Policy Association, 1975.

Borgstrom, Georg. *The Hungry Planet: The Modern World at the Edge of Famine.* New York: Macmillan Co., 1972.

———. *Too Many: A Study of the Earth's Biological Limitations.* New York: Macmillan Co., 1969.

Bridger, Gordon, and de Soissons, Maurice. *Famine in Retreat? The Fight Against Hunger: A Study and a Strategy.* London: J. M. Dent and Sons, 1970.

Brown, Lester R. *By Bread Alone.* New York: Praeger Publishers, 1974.

———. *Population and Affluence: Growing Pressures on World Food Resources.* Washington, D.C.: Overseas Development Council, 1973.

———. *Seeds of Change: The Green Revolution and Development in the 1970's.* New York: Praeger Publishers, 1970.

Cole, H.S.D.; Freeman, Christopher; Jahoda, Marie; and Pavitt, K.L.R. *Models of Doom.* New York: Universe Books, 1973.

Eckholm, Erik P. *Losing Ground: Environmental Stress and*

World Food Prospects. New York: W. W. Norton, 1976.

Gribbin, John. *Forecasts, Famines, and Freezes: Climate and Man's Future.* New York: Walker and Co., 1976.

Hailstones, Thomas J., and Mastrianna, Frank V. *Contemporary Economic Problems and Issues.* Cincinnati: South-Western Publishing Co., 1976.

Kahn, Herman; Brown, William; and Martel, Leon. *The Next 200 Years: A Scenario for America and the World.* New York: William Morrow and Co., 1976.

Keyfitz, Nathan. "United States and World Populations." In *Resources and Man: A Study and Recommendations.* San Francisco: W. H. Freeman, 1969.

Laffin, John. *The Hunger to Come.* London: Abelard-Schuman, 1971.

Laird, Roy D. "Soviet Agriculture in 1973 and Beyond in Light of United States Performance." In *The Soviet Union: The Seventies and Beyond.* Lexington, Mass.: D. C. Heath, 1975.

Meadows, Donella H.; Meadows, Dennis L.; Randers, Jorgen; and Behrens, William W. *The Limits to Growth.* New York: Universe Books, 1972.

Mesarovic, Mihajlo, and Pestel, Eduard. *Mankind at the Turning Point: The Second Report to the Club of Rome.* New York: E. P. Dutton, 1974.

Pimentel, David, et al. "Food Production and the Energy Crisis." In *Food: Politics, Economics, Nutrition, and Research.* Edited by Philip H. Abelson. Washington, D.C.: American Association for the Advancement of Science, 1975.

Ponte, Lowell. *The Cooling.* Englewood Cliffs, N. J.: Prentice-Hall, 1976.

Schneider, Stephen H. *The Genesis Strategy: Climate and Global Survival.* New York: Plenum Press, 1976.

Schneider, William. *Can We Avert Economic Warfare in Raw Materials? U.S. Agriculture as a Blue Chip.* New York: National Strategy Information Center, 1974.

——. *Food, Foreign Policy and Raw Materials Cartels.* New York: Crane, Russak, and Co., 1976.

Simon, Paul, and Simon, Arthur. *The Politics of World Hunger.* New York: Harper's Magazine Press, 1973.

Trezise, Philip H. *Rebuilding Grain Reserves—Toward an International System.* Washington, D.C.: The Brookings Institution, 1976.

Tydings, Joseph D. *Born to Starve.* New York: William Morrow and Co., 1970.

Waterlow, Charlotte. *Superpowers and Victims.* Englewood Cliffs, N. J.: Prentice-Hall, 1974.

Periodicals

Arthur, Henry B., and Cramer, Gail L. "Brighter Forecast for the World's Food Supply." *Harvard Business Review*, May-June 1976, pp. 161-66.

Bradsher, Henry S. "U.S. Experts See Soviet Grain Shortfall." *The Washington Star*, 17 November 1976, p. A-4.

Brink, R. A.; Densmore, J. W.; and Hill, G. A. "Soil Deterioration and the Growing World Demand for Food." *Science* 197 (August 1977):625-30.

Brown, Lester R. "The World Food Prospect." *Science* 190 (December 1975):1053-59.

Bryson, Reid A. "The Lessons of Climatic History." *Ecologist,* July 1976, pp. 205-11.

——. Interview in *Mother Earth News*, March 1976, pp. 7-17.

Bryson, Reid A., and Alexander, T. "Ominous Changes in the World's Weather." *Fortune*, February 1974, pp. 90-95.

Coale, Ansley J. "The History of the Human Population." *Scientific American*, September 1974, pp. 41-51.

Crosson, Pierre R. "Institutional Obstacles to Expansion of World Food Production." *Science,* 188 (May 1975):519-24.

Demeny, Paul. "The Populations of the Underdeveloped Countries." *Scientific American*, September 1974, pp. 149-59.

Dovring, Folke. "Soybeans." *Scientific American,* February 1974, pp. 14-21.

Freedman, Ronald, and Berelson, Bernard. "The Human Population." *Scientific American,* September 1974, pp. 31-39.

Frisch, Roger, and Revelle, Roger. "Nutrition and Childbearing Age." *Archives of Diseases of Childhood* 46, no. 249 (October 1971):695-701.

Grant, James P. "Food, Fertilizer, and the New Global Politics of Resource Scarcity." *The Annals of the American Society of Political and Social Science* 420 (July 1975):11.

Hall, Charles R. "Mobilizing the Multinational." *The Conference Board Record,* July 1975, pp. 47-48.

Heady, Earl O. "The Agriculture of the United States." *Scientific American,* September 1976, pp. 107-27.

Hopkins, Raymond F. "How to Make Food Work." *Foreign Policy,* Summer 1977, pp. 89-107.

Hopper, David. "The Development of Agriculture in Developing Countries." *Scientific American,* September 1976, pp. 196-205.

Latham, Michael C. "Nutrition and Infection in National Development." *Science* 188 (May 1975):561-65.

Matthews, Samuel W. "What's Happening to Our Climate?" *National Geographic,* November 1976, pp. 576-615.

Mayer, Jean. "The Dimensions of Human Hunger." *Scientific American,* September 1976, pp. 40-49.

Morgan, Dan. "Our Newest Weapon: Food." *Saturday Review,* 13 November 1976, pp. 7-8.

Nove, Alec. "Will Russia Ever Feed Itself?" *New York Times Magazine,* 1 February 1976, p. 9.

The OECD Observer (Paris), May-June 1976.

Poleman, Thomas J. "World Food: A Perspective." *Science* 188 (May 1975):510-18.

Randal, Judith. "Will Congress Do Something About the Weather?" *Change,* August 1976, p. 50.

Revelle, Roger. "Food and Population." *Scientific American,* September 1974, pp. 161-70.

———. "The Resources Available for Agriculture." *Scientific American,* September 1976, p. 165.

Ross, Douglas N. "Whatever Happened to the Green Revolution?" *The Conference Board Record,* 28 October 1976, pp. 56-59.

Rothschild, Emma. "Concocting the Next Crisis." *The New York Review of Books,* 4 April 1974, p. 30.

Sanderson, Fred H. "The Great Food Fumble," *Science* 182 (May 1975): 503-09.

Schertz, Lyle A. "World Food: Prices and the Poor." *Foreign Affairs* 52 (April 1974):511.

Schmid, Gregory. "Investment Opportunities in Food Production in LDC's." *Food Policy,* May 1976, pp. 220-31.

Shapley, Deborah. "Crops and Climatic Change: USDA's Forecasts Criticized." *Science* 193 (September 1976):1222-24.

Splinter, William E. "Center Post Irrigation." *Scientific American,* June 1976, pp. 90-99.

Steif, William. "OPEC Welshes on a Promise." *Saturday Review,* 13 November 1976, p. 14.

Steinhart, John S., and Steinhart, Carol E. "Energy Use in the U.S. Food System." *Science* 184 (May 1974):307.

Thompson, Louis M. "Weather Variability, Climate Change, and Grain Production." *Science* 188 (May 1975):535-41.

Varon, Bension, and Takeuchi, Kenji. "Developing Countries and Non-Fuel Minerals." *Foreign Affairs* 52 (April 1974): 497-510.

Walters, Harry. "Difficult Issues Underlying Food Problems." *Science* 188 (May 1975):524-30.

Westoff, Charles F. "The Populations of the Developing Countries." *Scientific American,* September 1974, pp. 109-20.

Willett, Joseph W. "Food as a Factor in U.S.-USSR Relations." *National Security Affairs Forum,* Spring-Summer 1976, pp. 33-37.

——. "The World Food Situation and Prospects: Optimism vs. Pessimism," *National Security Affairs Forum,* Fall-Winter 1975, pp. 17-24.

Wolfgang, Marvin E., ed. "Adjusting to Scarcity." *The Annals of the American Academy of Political and Social Science* vol. 420, July 1975.

Wortman, Sterling. "Food and Agriculture." *Scientific American,* September 1976, pp. 31-39.

Government Documents

National Academy of Sciences. *Climate and Food.* Washington, D.C., 1976.

———. *Understanding Climatic Changes, A Program for Action.* Washington, D.C., 1975.

National Academy of Sciences, Committee on Climate and Weather Fluctuations and Agricultural Production. *Climate and Food: Climatic Fluctuations and U.S. Agricultural Production.* Washington, D.C., 1976.

U.S. Agency for International Development. *The Population Explosion: A Present Danger.* Washington, D.C., 1975.

U.S. Central Intelligence Agency. *A Study of Climatological Research As It Pertains to Intelligence Problems.* Washington, D.C., 1974.

———. *Potential Implications of Trends in World Population, Food Production, and Climate.* Unclassified report no. OPR-401. Washington, D.C., 1974.

U.S. Congress, House. *Remarks by Representative George E. Brown on the National Climate Program Act of 1977.* In *Congressional Record,* 95th Cong., 1st sess., January 1977.

———. *Report on Data and Analysis Concerning the Possibility of a U.S. Food Embargo as a Response to the Present Arab Oil Boycott.* Report no. 33-674. Washington, D.C.: Government Printing Office, 1973.

———. *The International Development and Food Assistance Act of 1975.* H. Rept. 9005, Pub. L. 94-161, 94th Cong., 1st sess., 1975.

U.S. Congress, House, Committee on Foreign Affairs. *Report by Subcommittee on Foreign Economic Policy of the Committee on Foreign Affairs.* Washington, D.C.: Government Printing Office, 1974.

U.S. Congress, House, Committee on International Relations.

Hearings Before the Subcommittee on International Resources, Food, and Energy. Res. 393, 94th Cong., 2nd sess., June 1976.

U.S. Congress, Joint Economic Committee. *Impact of Russian Grain Purchases on Retail Food and Farm Prices and Farm Income in the 1975 Crop Year.* Prepared by George E. Brandow. Washington, D.C.: Government Printing Office, 1975.

——. *1976 Annual Report.* Washington, D.C.: Government Printing Office, 1976.

U.S. Department of Agriculture. *Handbook of Agricultural Charts, Agricultural Handbook No. 504.* Washington, D.C.: U.S. Department of Agriculture, 1976.

U.S. Department of Agriculture, Economic Research Service. "An Assessment of Government Programs that Protect Agricultural Producers from Natural Risks." Prepared by Thomas A. Miller and Alan S. Walter. In *Agricultural Food Policy Review.* Washington, D.C.: U.S. Department of Agriculture, 1977.

——. "Farmland, Will There Be Enough?" Prepared by M. L. Cotner and O. Krause. *U.S. Department of Agriculture Bulletin.* No. ERS-584. Washington, D.C., 1975.

——. "Livestock-Grain Interdependence: Implications for Policy." Prepared by Ronald A. Gustafson. In *Agricultural Food Policy Review.* No. ERS-AFPR-1. Washington, D.C.: U.S. Department of Agriculture, 1977.

——. *The World Food Situation and Prospects to 1985.* Foreign Agricultural Economic Report no. 98. Washington, D.C.: Government Printing Office, 1974.

U.S. Department of the Interior, Bureau of Mines. *U.S. Minerals Balance Sheet.* Washington, D.C.: U.S. Department of the Interior, 1975.

U.S. Department of State, Bureau of Public Affairs, Office of Media Services. "Global Consensus and Economic Development." Addendum by Henry A. Kissinger. In *Addendum to Seventh Special Session, UN 60, September 1, 1975.* Washington, D.C.: U.S. Department of State, 1975.

Other Documents

Climate, Change, Food Production and Interstate Conflict— A Bellagio Conference, 1975. New York: The Rockefeller Foundation, 1976.

Iowa State University. *The Iowa Stater,* 3, no. 7, 1977.

Koffsky, Nathan M. "Food Needs of the Developing Countries." Paper presented at USDA-sponsored conference on food policy at the Pan American Health Organization held in Washington, D.C., 23-29 April 1977.

1977 Policy Statement adopted by 35th Annual Convention of the Minnesota Farmer's Union held in St. Paul, Minnesota, 21-23 November 1976.

United Nations Working Paper No. 55. New York, 1975.

Webster, Charles P. *Population Growth: Major Threats to the Generation of Peace.* Washington, D.C.: Strategic Research Group, The National War College, 1973.

World Meteorological Organization. *The Physical Basis for Climate Modelling.* Global Atmospheric Research Program Pub. no. 16. Geneva, 1975.

Printed and bound by CPI Group (UK) Ltd, Croydon, CR0 4YY

23/10/2024

01778241-0015